MOLECULAR MECHANISMS INFLUENCING AGGRESSIVE BEHAVIOURS

The Novartis Foundation is an international scientific and educational charity (UK Registered Charity No. 313574). Known until September 1997 as the Ciba Foundation, it was established in 1947 by the CIBA company of Basle, which merged with Sandoz in 1996, to form Novartis. The Foundation operates independently in London under English trust law. It was formally opened on 22 June 1949.

The Foundation promotes the study and general knowledge of science and in particular encourages international co-operation in scientific research. To this end, it organizes internationally acclaimed meetings (typically eight symposia and allied open meetings and 15–20 discussion meetings each year) and publishes eight books per year featuring the presented papers and discussions from the symposia. Although primarily an operational rather than a grant-making foundation, it awards bursaries to young scientists to attend the symposia and afterwards work with one of the other participants.

The Foundation's headquarters at 41 Portland Place, London W1B 1BN, provide library facilities, open to graduates in science and allied disciplines. Media relations are fostered by regular press conferences and by articles prepared by the Foundation's Science Writer in Residence. The Foundation offers accommodation and meeting facilities to visiting scientists and their societies.

Information on all Foundation activities can be found at http://www.novartisfound.org.uk

Novartis Foundation Symposium 268

MOLECULAR MECHANISMS INFLUENCING AGGRESSIVE BEHAVIOURS

2005

John Wiley & Sons, Ltd

Published in 2005 by John Wiley & Sons Ltd,
The Atrium, Southern Gate,
Chichester PO19 8SQ, UK

National 01243 779777
International (+44) 1243 779777
e-mail (for orders and customer service enquiries): cs-books@wiley.co.uk
Visit our Home Page on http://eu.wiley.com

Other Wiley Editorial Offices

John Wiley & Sons Inc., 111 River Street, Hoboken, NJ 07030, USA

Jossey-Bass, 989 Market Street, San Francisco, CA 94103-1741, USA

Wiley-VCH Verlag GmbH, Boschstr. 12, D-69469 Weinheim, Germany

John Wiley & Sons Australia Ltd, 33 Park Road, Milton, Queensland 4064, Australia

John Wiley & Sons (Asia) Pte Ltd, 2 Clementi Loop #02-01, Jin Xing Distripark, Singapore 129809

John Wiley & Sons Canada Ltd, 22 Worcester Road, Etobicoke, Ontario, Canada M9W 1L1

Wiley also publishes its books in a variety of electronic formats. Some content that appears in print may not be available in electronic books.

Novartis Foundation Symposium 268
viii+264 pages, 22 figures, 5 tables

British Library Cataloguing in Publication Data

A catalogue record for this book is available from the British Library

ISBN-13 978-0-470-01068-6
ISBN-10 0-470-01068-1

Typeset in 10½ on 12½ pt Garamond by Dobbie Typesetting Limited, Tavistock, Devon.
Printed and bound in Great Britain by T. J. International Ltd, Padstow, Cornwall.
This book is printed on acid-free paper responsibly manufactured from sustainable forestry, in which at least two trees are planted for each one used for paper production.

Contents

Symposium on Molecular mechanisms influencing aggressive behaviours, held at the Novartis Foundation, London, 20–22 July 2004

Editors: Gregory Bock (Organizer) and Jamie Goode

This meeting was based on a proposal made by Donald Pfaff, Barry Keverne and Randy Nelson

Participants

D. Caroline Blanchard University of Hawaii, Department of Psychology, College of Social Sciences, Gartley 110, 2430 Campus Road, Honolulu, HI 96822, USA

Robert J. Blanchard University of Hawaii, Department of Psychology, College of Social Sciences, Gartley 110, 2430 Campus Road, Honolulu, HI 96822, USA

Björn Brembs Institut für Neurobiologie, F U Berlin, Königin-Luise-Str. 28/30, D-14195 Berlin, Germany

Edward S. Brodkin University of Pennsylvania School of Medicine, Center for Neurobiology and Behavior, 415 Curie Boulevard, Room 111, Philadelphia, PA 19104-6140, USA

Ian Craig PO Box 82, SGDP Research Centre, Institute of Psychiatry, De Crespigny Park, Denmark Hill, London SE5 8AF, UK

Catherine Dulac Department of Molecular and Cellular Biology, Harvard University, 16 Divinity Avenue, Cambridge, MA 02138, USA

Craig Ferris Department of Psychiatry, University of Massachusetts Medical School, 55 Lake Avenue North, Worcester, Massachusetts 01655, USA

Stephen Gammie 1117 West Johnson St, Zoology Research Building, Room 213, Department of Zoology, University of Wisconsin, Madison, WI 53706, USA

Robert Hinde St John's College, Cambridge CB2 1TP, UK

Barry Keverne Sub-Department of Animal Behaviour, University of Cambridge, High Street, Madingley, Cambridge CB3 8AA, UK

Jaap Koolhaas Department of Animal Physiology, University of Groningen, P.O. Box 14, 9750 AA Haren, The Netherlands

Klaus-Peter Lesch Klinik und Poliklinik für Psychiatrie und Psychotherapie, Universität Würzburg, D-97080 Würzburg, Germany

Robin Lovell-Badge The National Institute for Medical Research, The Ridgeway, Mill Hill, London NW7 1AA, UK

Stephen Manuck Behavioral Physiology Laboratory, Department of Psychology, University of Pittsburgh, 506 Old Engineering Hall, Pittsburgh, PA 15260, USA

Manuela Martinez Department of Psychobiology, Faculty of Psychology, University of Valencia, Avda. Blasco Ibañez 21, 46010, Valencia, Spain

Randy Nelson Departments of Psychology and Neuroscience, 09 Townshend Hall, The Ohio State University, Columbus, OH 43210, USA

Berend Olivier Department of Psychopharmacology, Utrecht Institute for Pharmaceutical Sciences, Utrecht University, Sorbonnelaan 16, 3584 CA, Utrecht, The Netherlands

Donald Pfaff (*Chair*) Neurobiology and Behavior, The Rockefeller University, Box 275, 1230 York Avenue, New York, NY 10021-6399, USA

Diane M. Robins Department of Human Genetics, University of Michigan, 4909 Buhl, 1241 E. Catherine St, Ann Arbor, MI 48109-0618, USA

David Skuse Behavioural and Brain Sciences Unit, Institute of Child Health, 30 Guilford Street, London WC1N 1EH, UK

Stephen Suomi National Institute of Child Health and Human Development, 31 Center Drive, Bethesda, MD 20892, USA

Brian Trainor (*Novartis Foundation Bursar*) Bauer Center for Genomics Research, Harvard University, 7 Divinity Avenue, Cambridge, MA 02138, USA

Introduction

Donald Pfaff

Neurobiology and Behavior, The Rockefeller University, Box 275, 1230 York Avenue, New York, NY 10021-6399, USA

In this introduction I'd like to state some questions that might deserve consideration during this meeting. To try to do so in a comprehensive way would be presumptuous of me. So I have, off the top of my head, listed some back of the envelope questions that might put some of our discussions in a historical context, and then I will ask Robert Hinde, Barry Keverne and Randy Nelson to add considerations that they might have.

My first question might be titled 'Beyond transcription'. Certainly, with respect to androgenic effects on aggressive behaviours, up until now it has been easiest to ascribe regulatory influences of hormones to transcriptional changes. This was natural for a neurobiologist to do because the chemistry of DNA is simpler than that of most other cellular compounds. Until 1944, when DNA was discovered to be the genetic substance, it was considered to be a stupid molecule—much too stupid to carry the genetic information. So my question is, to what extent can we envision facing the difficulties of dealing with RNA editing and protein chemistry in neurons as part of the business of life, with respect to molecular mechanisms underlying aggression?

Second, how are we going to analyse the indirect as well as the direct influence of genomic changes on aggressive behaviours? In 1941, working with *Neurospora*, Beadle and Tatum came up with data that supported the one gene/one enzyme concept. I would argue that up until a few years ago, the one gene one enzyme concept dominated functional genomics, but with respect to sex behaviours where we know an awful lot, I have been able to argue that we are always dealing with patterns of gene expression governing patterns of sociosexual behaviours. How do we apply this kind of thinking to aggression?

Finally, ethology is the science of behaviour. I for one aspire to the kind of precision and lawfulness that we find in the physical sciences. To what extent do the stereopathies of aggressive behaviour, and their obvious biological adaptiveness (at least in animals), provide one of the best subjects for us to prove that behavioural science can achieve the same kind of lawfulness and precision as the physical sciences? I'll now ask Robert Hinde what hopes or warnings he might have for us.

Hinde: I am a misfit at this meeting because I haven't worked below the skin for 40 years, let alone at the molecular level. But I am of the opinion that violence is perhaps the most important problem humans face. What I will be looking for is the following. Are we making a clear distinction between aggression (which is usually in human studies defined as the intentional affliction of harm on another individual), and aggressiveness (the propensity to be aggressive). One is an act, the other is a general propensity. Second, are we making a clear distinction between the various types of aggression? These include instrumental or felonious aggression, teasing aggression, revenge aggression and socially acceptable aggression that is condoned in a gang or small group but not in the general population, and so on. This approximates to what Donald Pfaff calls the 'form' of aggression, when he makes the distinction between testosterone-instigated and maternal aggression. During the papers I shall be thinking of whether it is the motor pattern that is being studied (and some of the subjects are chosen because of the stereotyped nature of mouse aggression which makes it easy to score) or is it the propensity to use violence with any sort of motivation, or is it motivation being studied in a particular context.

Keverne: We have a broad audience here of researchers studying subjects ranging from flies to humans. There is no doubt that many of the same genes cross all of these biological groups. It is important to bear in mind that with the cloning and sequencing of the human and mouse genome, we now know that both have approximately the same number of genes. Actually, humans have slightly fewer than the mouse, and the ones we lack are primarily in the olfactory system. We also know that there is 96% sequence synteny. Clearly, these similar genes and sequences are building very different phenotypes. It is important for us to take into account how different these phenotypes are, particularly with respect to the brain and brain development. One of the things that is different between the human brain and the small rodent brain is that our behaviour is not primarily directed by olfactory cues. Secondly, most of the development in our brains occurs in the postnatal environment, and it occurs at a time when we are learning the social cues of how to control and regulate our behaviour. Evolutionarily, this development of the brain is not insignificant when we try to transfer information from what is known in the fly or mouse to the human. At the transcriptional level of a given gene it is going to be remarkably similar, but in terms of what the genome is doing in building a phenotype, it is going to be different, and in terms of what humans can do with that phenotype is also very different. For example, much of human behaviour is to some extent emancipated from hormonal influences. Women do not need to have undertaken pregnancy and parturition to be good mothers. We don't wait until we get the hunger signals from hormones before we feed. We do our foraging in a supermarket in advance of hunger. Aggression is even more complex because it invariably needs a context. It is rarely spontaneous and usually secondary to other behaviours. We need to take this enormous complexity into account before we try to translate what a gene might be doing at the cellular or neural level, which will be remarkably similar among diverse organisms, to what it is doing at the systems level in terms of how the whole brain functions.

Nelson: Nikolaas Tinbergen stated this most precisely and clearly when he said that description must precede analysis. I think it will be important for us, when talking about molecular mechanisms, to clearly define what aspect of aggression we are talking about. We often consider aggression as a monolithic behaviour, but there are many different components, and the molecular mechanisms underlying the various components of aggression likely differ. We need to be clear whether we are talking about the motivation to aggress, or motor patterns, or contextual cues as we consider the molecular mechanisms underlying aggression.

Some suggestions for revitalizing aggression research

Robert J. Blanchard and D. Caroline Blanchard*[1]

*Department of Psychology, University of Hawaii, 2430 Campus Road, Honolulu, HI 96822, and *Pacific Biomedical Research Center, University of Hawaii, 1993 East West Road Honolulu, HI 96822, USA*

Abstract. Aggression research is moribund. Lack of research over the past two decades has left many issues. (1) Understanding varieties of agonistic behaviour in an ethological context: categories differing in behaviours, target sites and function include offence, defensive attack, and predation. Biological systems must be determined for each of these. (2) Insuring availability of ethologically valid laboratory models of agonistic behaviour and describing (possibly species-specific) standards for these. We shall present models and consider the problematic issue of biting. (3) Use of non-damaging behavioural markers that precede fights. These should be independently analysed, measured and verified as potential substitutes for biting attack. (4) Interaction between fear and offensive aggressive motivation systems must be understood in order to evaluate whether independent variable (e.g. pharmacological, genetic) effects involve a specific motivational system rather than reflecting changes in oppositional systems. (5) Knowledge of agonistic systems and their biological basis must be extended to humans, focusing on both normal aggression in each category, and the development of models of aggressive psychopathology. Placing aggression research in an ethological context and focusing on its biomedical relevance may help to counter forces suppressing this work.

2005 Molecular mechanisms influencing aggressive behaviours. Wiley, Chichester (Novartis Foundation Symposium 268) p 4–19

Basic research in aggression: a dying field?

It is a considerable irony that during the past 30 years, a period in which problems of interpersonal and group violence in real world settings have come more and more strongly to the attention of the public and the media, the relative attention of the scientific community to basic experimental research on aggression has sharply declined. This statement is based on data for research on the three rodent

[1]This paper was presented at the symposium by Robert J. Blanchard. All correspondence should be addressed to D. Caroline Blanchard.

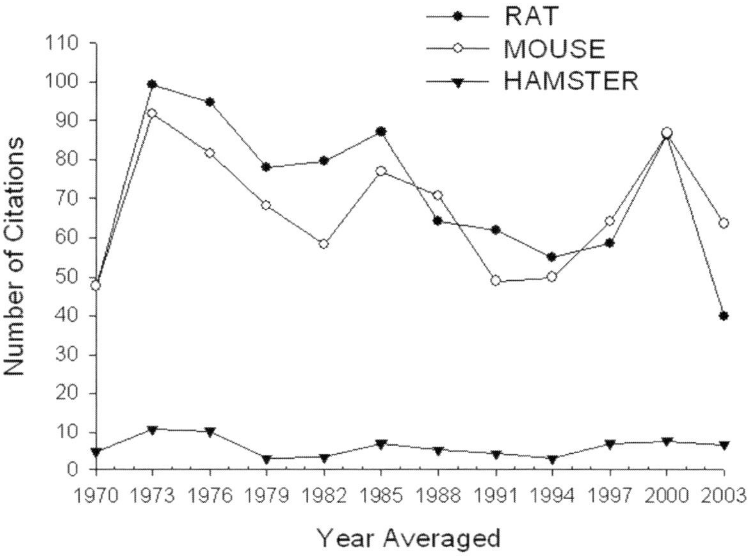

FIG. 1. Average number of citations per year from PubMed during successive three-year periods, retrieved in response to search terms 'aggressive behavior AND rat' ('. . . AND mouse' or '. . . AND hamster').

species that are used as subjects of most experimental aggression experiments: PubMed citations appearing in response to the search terms 'aggressive behaviour' and 'rat' (or mouse, or hamster) during three year periods from 1970 to 2000 indicate that citations for rats and mice are about equal, and appear to reflect a peak in the early 1970s followed by a slow decline. Hamster studies, fewer in number, remained relatively steady over this period.

In contrast to this steady state, comparable figures for studies of sexual and stress-related behaviour for the same three species of laboratory animals indicate a sharply increasing trend over the same period. Although both areas received attention comparable to that of aggression research in 1970, studies of sexual behaviour have since almost tripled, and stress-related behaviour increased nearly 20 times. Thus, on a baseline of activity in comparable areas of behavioural research, it can be seen that aggression research has substantially declined over the past 30 years. The comparison with stress research is particularly interesting, as increased restrictions on animal research involving aversive events might be blamed for the decline in experimental aggression work. However, the striking increase in rodent research on stress indicates that, although restrictions on animal research may have exerted considerable inhibitory influence overall, they

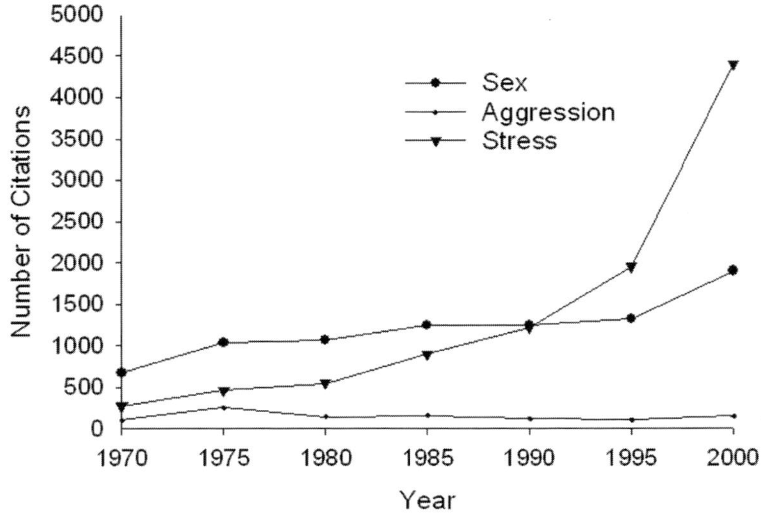

FIG. 2. Total number of citations per year for index years (1970, 1975, etc.) retrieved from PubMed in response to search terms 'stress AND rat' ('. . . AND mouse' or '. . . AND hamster' . . . summed over the three species), and for 'sexual behavior AND rat' ('. . . AND mouse' or '. . . AND hamster' . . . summed over the three species). These are shown in comparison to data for 'aggressive behavior AND. . . representing summed data over the three species from Fig. 1.

do not appear to have served as a specific and differential inhibitor of work involving response to stressful or provocative conditions.

Finally, while the absolute numbers of aggression studies involved may seem adequate even if suffering in comparison to other areas, these numbers provide an overestimation of the magnitude of true aggression research. Individual analysis of citations for a single year (2000) indicated that only about 24% of the studies retrieved by this search — some 36 studies for the year 2000 — contributed directly and empirically to our knowledge of offensive aggressive behaviour in the three laboratory species in which it has been best and most thoroughly analysed.

What is aggression, and why should it be studied?

Aggression is something of a 'catch-all' term for several types of evolved behaviours (Blanchard et al 1999). Like other evolved behaviour patterns, these behaviours are adaptive under a range of circumstances, but may be maladaptive in other situations. Analogously, evolved appetites for fats and sugars are adaptive when these nutrients are in relatively short or sporadic supply, but may result in widespread obesity and other health problems when such food items are abundant. Evolved behaviours also involve underlying brain and neurochemical systems, and additional problems may occur when these are hyperexpressed, resulting in pathological manifestation of behaviour; too much, too poorly controlled, or in the wrong situation (Marks & Nesse 1994). As will be detailed later, one variety of aggression — offensive aggression — is particularly sensitive to its own consequences, in the form of behaviour change on the part of the opponent. This sets up conditions for it to serve as an operant; offensive aggression increases when it is reinforced, leading directly to functions that are sometimes grouped under the rubric 'instrumental aggression' and providing for another avenue by which aggressive behaviour can become problematic. Finally, there is the problem of group aggression. This is a relatively rare phenomenon in other mammalian species, showing, however, a clear increase in larger-brained primates and reaching extraordinarily high levels in people. In combination with other factors such as the sensitivity of offensive aggression to successful consequences, and technological enhancement of the capacity of aggressive behaviour to cause damage, group aggression has been a prominent problem throughout human history (Keeley 1996).

This problem, that hyperexpressed and damaging aggression is common in human societies, constitutes one important reason for studying aggressive behaviour. The second is simply that aggression, even when normal and adaptive, includes several deeply interesting examples of evolved behaviour patterns that are simultaneously highly responsive to antecedent circumstances and to their own consequences. These patterns may be common across mammalian species, but there are important differences from one species to another in when, how, and to what effect the various types of aggression are expressed. All varieties of aggression known in inframammalian species are found in people, and the brain systems and behavioural budgets of both human and non-human mammals devote an extraordinary amount of space or time to them. They are worthy of study.

Aggression systems

In summary of material that is available in much greater detail elsewhere (e.g. Blanchard et al 1999), there are at least three major types of aggression in mammals: offensive aggression, defensive aggression and predatory aggression. Other categories, such as 'play fighting' or 'maternal aggression' have been proposed, but these either fit into one of the three rubrics given above (e.g. 'maternal aggression' may consist of both offensive and defensive attack components) or, as with play fighting, resemble aggression only superficially. Thus play fighting in rodents does not include hard bites, and it occurs in the context of amicable relations between the 'opponents' both pre- and post-encounter (Pellis 1988). Predatory aggression may be included under the rubric of aggression on the basis that it involves harm to the opponent. However, it is very different from both offensive and defensive attack in being aimed almost exclusively at non-conspecifics, and in being absent in many mammal species. While paradigms involving attacks by predators on animals of prey species, e.g. mouse- or frog-killing by rats, and insect-killing by mice (Karli 1956, Brain 1979) have sometimes been used to measure aggression, this is becoming increasingly rare. There is an emerging consensus that predation is very different from offensive and defensive attack on grounds of its core motivations and its relationship to aversive emotions and emotional arousal (see Blanchard et al 1999 for review), not to mention that human predation on prey animals is seldom regarded as constituting a social problem. For all these reasons, the study of predation is seldom grouped with work on other forms of aggression.

Offensive and defensive aggression

These comments will focus on offensive and defensive aggression, as reflecting two fundamental divisions of 'serious' aggression. The distinction between offensive and defensive aggression is based on a number of aspects of behaviour, including antecedent conditions, organismic variables, response topography and typical outcomes (Blanchard & Blanchard 1977, Brain 1979, Blanchard et al 1999). In terms of antecedents, offensive aggression involves response to challenge over adaptively important resources, whereas defensive aggression is attack in defence of the subject's own bodily integrity. Defensive attack may be seen to either attacking non-conspecifics, typically predators, or to attacking conspecifics. It is embedded in a larger pattern of defensive behaviours and, depending on species and circumstances, may or may not occur in a particular instance of defence against attack. It thus constitutes one component of a larger defence pattern, whereas offensive aggression stands alone.

With reference to the behaviours involved, within-species offensive and defensive aggression involve potentially damaging attack targeted towards specific areas on the body of a conspecific opponent. The targets for these two types of aggression are different. For rats and mice, conspecific offensive attack is targeted toward the back and flanks of the opponent while defensive attack is targeted at the snouts of both conspecifics and predators (Blanchard et al 1977a, 1980, Blanchard & Blanchard 1977). These patterns appear to be quite similar in wild and laboratory rats and mice, although some non-attack components of the defence pattern (notably freezing) have been altered by domestication (see Blanchard 1997 for review). In hamsters, the rump (lower back) and lower flanks are the targets of offensive attack (Pellis & Pellis 1988), but targets for defensive

TABLE 1 Target sites for conspecific offensive attack

Attack site (%)	*Rat*	*Mouse*	*Hamster*
Head	7.24	1.82	0.21
Back	88.62	78.67	78.27
Ventrum	0.00	10.29	19.65
Limbs	4.14	8.88	1.35
Genitals	0.00	0.00	0.52
Tail	0.00	0.34	0.00

attack have not been described. All of these targets apply to unanaesthetized conspecific opponents, animals that are able to display behavioural defences, and the defences of the opponent may have a considerable influence on target sites for attack (Blanchard et al 1977b).

These target sites are important for an understanding of both conspecific offensive aggressive and conspecific defensive behaviours, in that conspecific defensive behaviours are organized to protect the target site for offensive attack by making it unavailable for biting; in turn these defences strongly influence the offensive tactics used to gain access to these sites. In rats and mice defence against attacking conspecifics includes flight, freezing (a lot in laboratory rats, some in mice), manoeuvres to protect the specific targets of offensive attack, and defensive threat and attack. The target site-protecting manoeuvres include postures in which the defender faces the attacker, often in an upright defensive stance from which it may pivot easily, to continue to remove its back and flanks from the attacker, and lying on the back. Offensive attack behaviours consist of movements enabling the attacker to thwart these defensive behaviours, such as a 'lateral' approach to an upright defender, that involves moving forward and around it to bite at the flanks and back (Blanchard & Blanchard 1977, Blanchard et al 1979, Pellis & Pellis 1988).

The problem of 'animal cruelty': bites and wounds

The dynamics of these patterns make it comparatively easy to differentiate offensive and defensive attack in laboratory rodents on a purely behavioural basis, without prior knowledge of the antecedent conditions and history of the animal. They may make it possible to differentiate these patterns, and even to measure the intensity of attack, without permitting the actual damaging component of attack, the bite, to occur. Can offensive aggression be measured through ancillary behaviours such as piloerection of the subject or spacing by the opponent, or defensive attack by its accompanying threat behaviours (defensive upright postures, defensive vocalizations)? Animal research on evaluation of aggression by analysis of non-damaging elements is very much needed, as public abhorrence of what is seen as animal cruelty is an important reason for reductions in experimental work in this area.

An additional contribution to the problem of perception of 'cruelty' may include analysis and dissemination of information on the actual consequences of 'bites'. These represent the only component of attack in rodents (and most other mammalian species) that is capable of producing substantial tissue damage to the opponent. However, bites are not wounds. In recent hamster studies we noted that bites, measured directly by observation, were much more common than were actual wounds, evaluated on the same day, after the animal was sacrificed and completely denuded of hair. A follow-up study in mice confirmed this, that only about 1 bite in 10 actually cuts or tears the skin. The others appear to constitute a pinch — doubtless startling and sometimes painful, but different than what is commonly accepted as the consequence of being bitten. A more appropriate label for these, when cuts are not involved, might be 'toothpinch' or 'pinch-vocalization' when the defending animal vocalizes in response.

Aggressive and defensive motivations: unravelling a complex interaction

As an adaptive behaviour, aggression is facilitated by circumstances predicting that it will be successful (e.g. location in subject's own territory, weak opponent); and inhibited when it is likely that the animal will lose (e.g. following defeat: Scott 1966, Huhman et al 2003). Losing involves a strong potential for body harm, and the emotional response to this possibility, defensiveness, exerts a strong inhibitory influence on the expression of offensive aggression. Because defence is a particularly salient neurobehavioural system in rodents, genetic, pharmacological and experiential manipulations that alter offensive attack may work through defence as easily, and with as much potency, as through a direct effect on aggression itself. (Notably, this is not true of defensive aggression, e.g. Blanchard et al 1980, potentially providing an additional means of differentiating the two.)

These effects on defence may come from many sources. In addition to conspecific defeat, and response to predators or other threat features of the situation (e.g. novelty), defensiveness can be changed by genetic or pharmacological manipulations, strongly but indirectly impacting offensive aggression. Individual differences factors impacting human aggression may also operate through inhibitory systems (Brennan et al 1997). The effect of defensiveness on offensive aggression may be very situation-specific, depending on the level of threat initially engendered by the situation and opponent; the capability of the opponent to retaliate; and whether general defensiveness or specific aspects of defence have been enhanced or reduced by relevant manipulations. If the situation or manipulations used produce defensive attack, there is also the question of differentiating this from offensive aggression. At present, there are relatively few guidelines for such analyses, although studies specifically directed at analysis of antiaggression drugs have at least recognized some of the problems involved (e.g. Olivier et al 1994).

The problem of fear–aggression interactions is particularly damaging in view of a strategy increasingly used to avoid animal cruelty issues — measurement of aggression by a single measure, latency to attack. The behavioural inhibition associated with enhanced defence is virtually guaranteed to enhance latency to first attack. One analytic strategy may be to measure other fear-affected latencies, e.g. to copulation with a strange but receptive female, or eating in a novel situation. If they also increase, increased latency to attack cannot be interpreted as reduced aggressiveness. However, this is only a partial solution, and aggression research badly needs an algorithm for differentiating effects on aggressive motivations from changes in defensiveness. While it is probably too early to think about standardizing animal aggression protocols, a less formalized group of 'attention points' for designing and analysing such studies would be useful.

The fit between animal and human aggression

While attention to the above issues would undoubtedly improve animal aggression research, there is a remaining and substantial problem concerning its relationship to aggression in people. Animal–human behavioural correspondences always involve tricky issues, but these are particularly acute in the case of aggression. First, there have been a lot of exaggerated claims about the degree to which animal findings are relevant to people, creating a poor atmosphere in which naïve acceptance of such claims is mixed with suspicion regarding the relevance of all such findings. Unfortunately, this situation is reinforced by a lack of communication between researchers of animal and human aggression phenomena.

It should and must be acknowledged that, currently, there is only a limited range of human phenomena to which results of animal aggression research can be directly, legitimately and meaningfully applied. This does not mean that the potential for finding cross-species similarities in aggression is especially limited, but simply that we don't have an adequate data/analysis basis for evaluating them. What is needed — in addition to a resurgence of both laboratory and field work on aggression in animals — is, first, more research into the dynamics of human aggression and, second, research bridging the human–non-human mammal gap, to show both the extent and the limitations of using animal models to explain human aggression.

In addition to scientific rigour and creativity, these attempts will also involve a need for great sensitivity on the part of the research community. Aggression is a stigmatized activity, but intentions to control it are regarded with deep suspicion, particularly in the USA with its explicit emphasis on individuality and civil liberties. This constitutes double jeopardy for the researcher trying to understand aggression and to gain some insights into its legitimate and illegitimate functions in contemporary society. While a daunting picture, it nicely illustrates the point that the opposition to research is often directly proportional to its importance. Aggression research is important, and we need to get back to it.

References

Blanchard DC 1997 Stimulus and environmental control of defensive behaviours. In: Bouton M, Fanselow M (eds) The functional behaviorism of Robert C. Bolles: learning, motivation and cognition. American Psychological Association, Washington DC, p 283–305
Blanchard RJ, Blanchard DC 1977 Aggressive behavior in the rat. Behav Biol 21:197–224
Blanchard RJ, Blanchard DC, Takahashi T, Kelley M 1977a Attack and defensive behavior in the albino rat. Anim Behav 25:622–634
Blanchard RJ, Takahashi LK, Fukunaga KK, Blanchard DC 1977b Functions of the vibrissae in the defensive and aggressive behavior of the rat. Aggress Behav 3:231–240
Blanchard RJ, O'Donnell V, Blanchard DC 1979 Attack and defensive behaviors in the albino mouse (Mus musculus). Aggress Behav 5:341–352
Blanchard RJ, Kleinschmidt CF, Fukunaga-Stinson C, Blanchard DC 1980 Defensive attack behavior in male and female rats. Anim Learn Behav 8:177–183
Blanchard DC, Hebert MA, Blanchard RJ 1999 Continuity vs (political) correctness: animal models and human aggression. In: Haug M, Whalen R (eds) Animal models of human psychopathology. American Psychological Association, Washington DC, p 297–316
Brain PF 1979 Differentiating types of attack and defense in rodents. In: Brain PF, Benton D (eds) Multidisciplinary approaches to aggression research. Elsevier/North Holland, Amsterdam, p 53–77
Brennan PA, Raine A, Schulsinger F et al 1997 Psychophysiological protective factors for male subjects at high risk for criminal behavior. Am J Psychiatry 154:853–855
Huhman KL, Solomon MB, Janicki M et al 2003 Conditioned defeat in male and female syrian hamsters. Horm Behav 44:293–299
Karli P 1956 The Norway rat's killing response to the white mouse: an experimental analaysis. Behaviour 10:81–103
Keeley L 1996 War before civilization. Oxford University Press, Oxford
Marks IM, Nesse RM 1994 Fear and fitness: an evolutionary analysis of anxiety disorders. Ethol Sociobiol 15:247–261
Olivier B, Mos J, Raghoebar M, de Koning P, Mak M 1994 Serenics. Prog Drug Res 42:167–308
Pellis SM 1988 Agonistic versus amicable targets of attack and defense: consequences for the origin, function, and descriptive classification of play-fighting. Aggress Behav 14:85–104
Pellis SM, Pellis VC 1988 Play-fighting in the Syrian golden hamster Mesocricetus auratus Waterhouse, and its relationship to serious fighting during postweaning development. Dev Psychobiol 21:323–337
Scott JP 1966 Agonistic behavior of mice and rats: a review. Am Zool 6:683–701

DISCUSSION

Keverne: You showed data illustrating the area of the animal's back which was attacked: were these proportions or absolute numbers?

R Blanchard: Proportions.

Keverne: So did the wild-type animals show the same amount of aggression in total? Were there more attacks by the intruder rather than a different amount of total aggression?

R Blanchard: The surprising thing is that in terms of offensive aggression, when we work with wild rats we see roughly equivalent levels of attack behaviour as we do with laboratory Long Evans rats.

Keverne: On one of your charts there weren't any data for wild-caught animals. Were there no attacks on the head? If so, what does this tell you about alpha-rating males in the wild?

R Blanchard: In the snap-trapped animals we saw head wounds on some, but only a few, and they far fewer than the number of back wounds. This distribution is similar to what we get in groups in laboratory settings, and suggests that the same sort of thing occurs in the wild as in the lab.

C Blanchard: I should mention that these animals were trapped in sugar cane fields in Hawaii. This is fertile breeding ground for wild rats. Every two or three years these fields are burned before the sugar is harvested. These poor rats have no place to go when the fire comes so they run into adjacent areas, and the places we trapped were in the adjacent areas right after the burning took place. These animals would have done a lot more fighting than normal because in their flight they would have run into someone else's territory. Some proportion of the trapped animals would have been the territorial rats and others the invaders. We assume that animals with largely head wounds are likely to have been the territory holders, and those with back wounds the intruders. Overall, however, the head and back wound proportions are fairly close to what you get in laboratory colony situations.

Hinde: You rather dismissed play fighting, and it comes outside your definition of aggression because it isn't 'intended' to hurt. But it is a way of isolating parts of the motor patterns of fighting from the other sorts of motivational bases. It might be interesting from this point of view.

R Blanchard: Indeed it might. The work relevant here is that of Sergio Pellis (Pellis & Iwaniuk 2004) who has been looking precisely at the motor patterns involved in play, and their development. He has looked at their relationship to adult sexual and aggressive behaviour. At least for some species he suggests that the motor systems that seem to be prepared more heavily in play fighting may actually be sexual.

Suomi: The situation may be different in primates. It is premature to dismiss it across all species. Along the same lines, there is another form of behaviour for which the term 'aggression' has been tagged on, and this is self-injurious behaviour or self-aggression. There the distinction between biting and wounding may be as relevant as in the case you described here.

R Blanchard: One thing we need to understand more clearly is that biting may be inhibited. Whenever a male bites a female who is not receptive, for example, we see clear biting but never wounding. Often biting is inhibited. If we could measure the intensity and the severity of the bite without looking for lesions this would be helpful. I know that in play fighting, certainly, there is inhibition of biting which changes to more injurious biting at a certain age. I accept that it is wrong to totally eliminate play fighting in all species.

Nelson: In terms of developing alternative descriptors, do you know of a lesion or drug that would take out piloerection but not affect attack, for example?

R Blanchard: I should know, but I don't. My concern about piloerection is that we are only just developing techniques of measuring it more precisely with our cameras. When I was working with super 8 mm film or standard videotape, piloerection was very difficult to score. We could do it only by live scoring.

Olivier: I have never seen a drug that wipes out piloerection and allows the animal to still perform aggressive behaviour. It doesn't seem to be a pure autonomic response.

Martinez: You proposed ways of finding alternative models for studying aggression in animals. In humans obviously it isn't possible to put two people together in order to let them fight and kill each other. So we find a way of obtaining some information, using paradigms in which individuals can display their readiness to behave aggressively without real aggression. Do you think this would be possible in animals?

R Blanchard: For example, could I develop a model of aggressive arousal by the use of piloerection? It might be possible, and this is something that is being worked on. Is it going to be the same thing as a real fight? I don't know. One is the question of arousal; the other also involves the motivation, motor patterns and so on. We are always going to have to bring the two into some relationship. We'd like to do this in the human work, too. Doing human ethology of real aggression is one of the most challenging problems and has rarely been done.

Koolhaas: It occurs to me that part of the problem with society is due to the very old definition of aggression, which is inflicting harm on another. This neglects the idea that aggression is a highly functional form of communication with strong inhibitory control for the adverse effects of wounding. Using aggressive behaviour, animals are trying to tell each other something. In my view, we have to make a distinction between the functional form of aggression (as a form of social communication) and violence, which can be considered as the pathological form of uncontrolled aggression. The human studies are mainly involved in violence, the pathology. However, most of the animal studies are aimed at the functional social communication form.

Olivier: When we tried to develop these anti-aggressive drugs in the 1970s and 80s we thought that there were systems in the brain involved in aggression, and pathology in humans could be due to excessive activity in these systems. We tried to get this concept into human aggression research, but this failed. We wanted to develop drugs for treatment and this failed because psychiatrists can't believe that aggression is a disease. We had to go for depression or anxiety instead. Do you think that in humans aggression and violence are due to the activation of specific neural systems or are they secondary effects? This is the question we never solved.

R Blanchard: The answer is both, but I am committed to the notion that there is a neurobehavioural system for offence which in humans reflects the same sociobehavioural systems in anger. I certainly think that there is a neurobehavioural system for anger. The issue concerns what psychiatric syndromes we have involving anger dis-control. Depression is one.

Olivier: Is it depression which steers aggression or the other way round?

R Blanchard: I don't know. The trouble with the current techniques for psychiatric diagnosis is that they largely focus on syndromes rather than symptoms. It is therefore difficult to parcel this out. We need a better characterization of the specific psychopathologies that are related to anger. This is one of the problems that are understood at the NIH. We need a symptom-based rather than syndrome-based description of psychopathology.

Skuse: I am a child psychiatrist, and I am interested in aggression in relation to social mis-perception. Many children we see who are aggressive appear to show these aggressive responses because they mis-perceive cues that other children are displaying. These might be facial expressions. If someone is looking angry when you are talking to them most of us would probably recognize that we should back off. Among children with autism spectrum disorders in particular, we find the perception of these cues is impaired. This is often how they get into fights. There is not necessarily any innate aggression in many of these encounters, but the situation can escalate to the point that the child has to be removed from a social situation, such as school. Preliminary work of ours suggests that this problem could be so widespread that it underlies many cases of so-called 'conduct disorder' (Gilmour et al 2004). This is one of those wonderful psychiatric definitions of a 'syndrome' that is actually a description of a collection of behaviours, which may have a heterogeneous aetiology.

Suomi: In many of the primate groups we have been observing we have been struck by what appears to be social ineptitude displayed by certain individuals. They apparently are not very good at perceiving social cues and responding appropriately when they receive feedback in a potentially ambiguous situation. Their behavioural output would be described as aggressiveness, but rather than consider it to be an innate quality, it seems to be more a lack of social recognition and an inability to read signals that other individuals read routinely and accurately.

Skuse: David Amaral at University of California Davis has also been studying social misperception in primates (Amaral & Corbett 2003). He is particularly interested in amygdalectomized animals, and the way that this affects their social behaviour.

Pfaff: In certain streets in New York, eye contact held for milliseconds too long can lead to murder among 'normal' individuals.

Skuse: Eye contact is fascinating. It appears to be a very potent stimulus for alerting this amygdala-related circuit that David Amaral is so interested in (Morris et al 2002). We have discovered during studies of normal individuals by functional magnetic resonance imaging (fMRI) that one can get a threat response, as measured by increased skin conductance, just by showing stimuli that are equivalent to eyes, even if these are not eyes that are in a face. Two dots in an abstracted form of threat cue will provoke such a response (J. Morris & D. Skuse, unpublished data). Fear recognition, from another's face, appears to predict differences in social cognitive competence, as least in males, and has predictable neural correlates (Corden et al 2005). There may be subcortical pathways (Pasley et al 2004) that mediate such responses in humans that are normally controlled by prefrontal cortical mechanisms (Holland & Gallagher 2004). Cortical mechanisms can inhibit this response rather rapidly after it occurs. If this inhibition should be compromised for some reason (for example, because of alcohol intoxication), then inappropriate aggressive (in the sense of defensive) behaviour may result.

Brodkin: IACUCs (Institutional Animal Care and Use Committees) are understandably concerned about injury to animals during the course of aggression testing, although you've presented data demonstrating that injury is rare. But I wonder whether studies of 'pre-bite' behaviours alone (e.g. piloerection, threat behaviours, or tail rattling) can serve as adequate substitutes for studying bite behaviours. Is it possible that these pre-aggressive threat behaviours may have a somewhat different underlying biology from the actual bite behaviours? Might there be certain individuals whose 'bark is worse than their bite'—who do a lot of threatening but don't progress to attack?

R Blanchard: You have hit on a major point. We are working on systems that are oppositional. The defence systems are opposing the offence systems. The two are always in conflict. If I want to use piloerection as a measure of aggression, I have to use it in an animal that is highly experienced and is clearly primarily offensive and not showing very much in terms of defensive behaviour. Such measures have to be used cautiously.

C Blanchard: If you pre-test Long Evans rats for aggressiveness and then put them all together in a visible burrow system with females, they quickly start fighting and form a dominance hierarchy within the first couple of days. There is little or no relationship between the latency to first attack and who becomes dominant. The one who *persists* in the individual pretests is usually the one who in a visible burrow system becomes dominant. Thus there are going to be some real problems translating the preattack measures, such as latency to first attack, into other more comprehensive measures of aggression. However, the value of such preattack measures also depends on what you are using them for. Perhaps you are not interested in who is going to become dominant; you are just interested in this first, immediate system and how it may interact with fear or behavioural inhibition.

Brodkin: Some IACUCs allow observation of pre-bite behaviours and one or two bites, but are less comfortable with allowing the aggressive behaviour to go on past the initial attack. But it seems to me that the transition from threat to the initiation of attack (e.g. first bite) is important and very much worthy of study. In clinical psychiatry, patients may get angry and may threaten, but the transition from threatening behaviour to actual harming behaviour of self or others is critically important.

Pfaff: With hamsters the first bite is crucial in terms of the resulting ethogram.

R *Blanchard:* The first bite is critical, but if you let the encounter run for a bit, you may see that some drugs affect the latency of bite but may not primarily be doing this through offence; they may be acting on defensive behaviour. This is frequently our problem with the initial stage measures.

Moreover, we do feel that some further attack behaviours can be permitted without abuse or cruelty. For example, we need to explain to IACUCs that what is frequently termed a 'bite' is actually measured when one animal contacts another with its snout and the latter vocalizes. In reality most of these 'pinch/vocalizations' do not result in wounds. IACUCs tend to equate the 'bite' measure with bloody wounds, and this is far from the truth. We are suggesting that conspecific aggression is an evolutionary punishment rather than a murderous social strategy.

Ferris: The communication component in the initiation of violence is extremely important. We illustrated this in a simple hamster model. Hamsters rapidly develop dominant–subordinate relationships with a minimum of attacks. Initially, there is intense overt aggression characterized by bites and attacks; however, on subsequent encounters there is little if any fighting. For months afterwards they communicate their social status with no aggression. They communicate through a behaviour called flank marking. They have pheromone-producing glands on their sides that they use to scent mark their environment as a form of olfactory communication. If you remove the flank glands from either the dominant or submissive animal they are unable to communicate. Consequently, whenever they encounter each other they fight. The outcome is always the same — dominant animals do all of the biting and attacking while the submissive animals try to run away. Once the lines of communication are broken the level of aggression is elevated. This would appear to be true in humans. I canvassed psychiatrists, asking them about where aggression and violence fit in serious mental illness, and most everyone agrees it is secondary to a primary mental illness. As you move across the spectrum of normal behaviour to severe mental illness where do you cross the line? To the clinicians, the tell-tale signs are agitation, impulsivity and the inability to communicate. In this meeting we should keep this cognitive component of communication in the mix, as we try to translate from the animal models to humans.

Pfaff: Applying your thinking to reproductive physiology, hamsters are a species where a non-receptive female can beat up a male to the point of killing him. How many other mammalian species are like this? What about dominance in rhesus monkeys?

Suomi: It depends on the situation, but there are instances where young males who seem to be inappropriately aggressive are targeted by high-ranking adult females, with the apparent intent of driving those young males out of their social group prematurely. These females are often successful. In some cases they kill the young male in the process.

Koolhaas: The approach we take now is to try to define the rules of the game. Aggressive interactions are heavily guided by rules. By analysing action–reaction patterns we aim at defining when the behaviour becomes pathological. We try to identify which individuals, in which situations, don't play the rules of the game anymore. We have individuals who are inclined to develop violent types of aggressive behaviour in which they no longer respond to (submission) signals from the opponent. I think the pathological form is completely different from the normal functional social communication patterns. The task is to find the rules of the game.

Manuck: There is some recent work on Intermittent Explosive Disorder (IED) related to social information processing in aggressive behaviour that may be relevant here (Best et al 2002). IED patients exhibit impaired recognition of facial expressions conveying emotion and show a strong preference for immediate (over delayed) rewards in simulated gambling experiments rigged to render such choices disadvantageous. One speculation is that these deficits arise from dysfunction of inhibitory connections between orbitomedial prefrontal cortex and the amygdala. Since there are two components — impulsivity and a failure to read social cues — do these patients show primarily an inability to process socially relevant information, or are they just too impulsive to do so in certain circumstances?

Skuse: The ability accurately to interpret these cues is distributed non-normally in the general population (Lawrence et al 2005). There are many of us who are pretty poor at it. If we also happen to have impulse control problems, we could be vulnerable to these sorts of outbursts. However, impulse control and the ability to read social signals are two different processes, and deficits in them do not necessarily coincide (Geurts et al 2004).

Craig: Has ADHD been looked at in terms of the inability to perceive social cues?

Skuse: Not as systematically (e.g. Cadesky et al 2000). Even in autistic children, the work has until now been confined to rather small groups, many of whom have had learning difficulties. Studies of children with autism with normal-range intelligence are lacking. Where it is pretty clear that they do have problems reading social cues, the exact nature of these problems at a neural level, and the extent of their impairment, is not yet known although progress is being made (Insel & Fernald 2004).

Pfaff: James Swanson at the University of California Irvine has studied large samples of ADHD children. He says that in boys but not in girls there is a large degree of co-morbidity between ADHD and other behavioural symptoms that have to do with aggression.

References

Amaral DG, Corbett BA 2003 The amygdala, autism and anxiety. In: Autism: neural basis and treatment possibilities. Wiley, Chichester (Novartis Found Symp 251) p 177–197

Best M, Williams JM, Coccaro EF 2002 Evidence for a dysfunctional prefrontal circuit in patients with an impulsive aggressive disorder. Proc Natl Acad Sci USA 99: 8448–8453

Cadesky EB, Mota VL, Schachar RJ 2000 Beyond words: how do children with ADHD and/or conduct problems process nonverbal information about affect? J Am Acad Child Adolesc Psychiatry 39:1160–1167

Corden B, Critchley HD, Skuse DH, Dolan RJ 2005 Fear recognition ability predicts individual differences in social cognition and neural functioning. Submitted

Geurts HM, Verte S, Oosterlaan J et al 2004 Can the Children's Communication Checklist differentiate between children with autism, children with ADHD, and normal controls? J Child Psychol Psychiatry 45:1437–1453

Gilmour J, Hill B, Place M, Skuse DH 2004 Social communication deficits in conduct disorder: a clinical and community survey. J Child Psychol Psychiatry 45:967–978

Holland PC, Gallagher M 2004 Amygdala-frontal interactions and reward expectancy. Curr Opin Neurobiol 14:148–155

Insel TR, Fernald RD 2004 How the brain processes social information: searching for the social brain. Annu Rev Neurosci 27:697–722

Lawrence K, Bernstein D, Pearson R, Mandy W, Skuse D 2005 How good are children and adolescents at recognizing facial expressions of emotion? A normative study. Submitted

Morris JS, deBonis M, Dolan RJ 2002 Human amygdala responses to fearful eyes. Neuroimage 17:214–222

Pasley BN, Mayes LC, Schultz RT 2004 Subcortical discrimination of unperceived objects during binocular rivalry. Neuron 42:163–172

Pellis SM, Iwaniuk AN 2004 Evolving a playful brain: a levels of control approach. Int J Comp Psychol 17:92–118

Aggressive behaviour: contributions from genes on the Y chromosome

Robin Lovell-Badge

Division of Developmental Genetics, MRC National Institute for Medical Research, The Ridgeway, Mill Hill, London NW7 1AA, UK

Abstract. The Y-linked gene *SRY* initiates male development in mammals through a pathway of gene activity leading to testis development. The testis then exports the male signal to the rest of the embryo via secreted molecules such as testosterone. These subsequently lead to many male characteristics in terms of anatomy, physiology and behaviour. Because males tend to be more aggressive than females, and androgens are often blamed for this, *SRY* can be thought of as a contributor to such behaviour. However, any effect the gene has is very indirect. Nevertheless, accumulating evidence suggests that other sex-linked genes may have more direct effects on differences between the sexes, and some of these are likely to include behavioural phenotypes. While it is not yet clear how important these are, they could compound decisions when treating cases of sex reversal and intersex conditions. They should also be borne in mind as a source of genetic variation when looking at differences between individuals, including behavioural traits such as aggression.

2005 Molecular mechanisms influencing aggressive behaviours. Wiley, Chichester (Novartis Foundation Symposium 268) p 20–41

Sex determination in mammals depends on the presence or absence of the Y chromosomal gene *Sry* (Koopman et al 1991, Lovell-Badge et al 2002). This acts in a dominant way to induce the differentiation of testes rather than ovaries from the indifferent gonad. Subsequently, male and female mammals differ in many aspects of anatomy, physiology and behaviour. The latter includes different types and degrees of aggression, in addition to reproductive and other social behaviours. But how do the differences between the sexes arise and what are the critical factors that can influence behaviour in a sex-specific manner?

Experiments conducted by Alfred Jost in the 1940s provided the first clear answers (Jost 1953). He was able to remove the early gonads from rabbit embryos *in utero* and then allow development to proceed. In this way he showed that embryos without ovaries continued developing as females, whereas embryos from which the early testes had been removed no longer gave rise to normal males but frequently developed as females instead. These findings have been replicated more recently by genetic means. For example, mouse embryos homozygous for a targeted null mutation in *Sf1*, a transcription factor expressed in the very early gonad, fail to develop either testes or ovaries, and in both cases they subsequently develop as females (Luo et al 1994).

The first, somewhat retrospective conclusion derived from these results is that genes initiating male development must act specifically within the developing gonad; they do not need to and are unlikely to function in all cells of the embryo. This is different from the situation in animals such as *Caenorhabditis elegans* and *Drosophila*, where the sex determining genes operate from very early stages and in all cells. This conclusion fits with what we now know about *Sry* (see below). The second conclusion, made by Jost, and shown by his subsequent work and that of others, was that the embryonic testes must make factors that trigger male development in the rest of the embryo.

One dogma also arose from these results, namely that all differences between males and females are due to these factors. Jost almost certainly did not look at all sex-specific traits in his rabbits, especially behaviour, as he would have been interested only in morphology, so it is not clear where this dogma came from. In theory it would be possible to look at behaviour in mice where gonads have been lost through mutation of *Sf1*, however, this also has a role in adrenal and pituitary function (Luo et al 1994). Although many differences are a consequence of factors secreted by the testis, it is now clear that they do not explain all male characteristics, including some behaviours. Evidence for this and the underlying causes are discussed below.

We know of three factors produced by the embryonic testis that act at a distance and have masculinizing effects on the rest of the embryo:

- Anti-Mullerian hormone (AMH, otherwise known as Mullerian inhibiting substance, MIS) is a member of the transforming growth factor (TGF)β superfamily of growth factors (Josso & Picard 1986, Munsterberg & Lovell-Badge 1991). This is made by Sertoli cells as soon as they begin to differentiate in the testis and has, as its main role, the function to eliminate the anlagen of the female reproductive tract from male embryos. In the absence of AMH, the Mullerian (or paramesonephric) ducts give rise to the oviducts, uterus, cervix and upper vagina.
- Testosterone begins to be secreted by Leydig cells in the fetal testis as soon as they differentiate (Morohashi 1997). This has many roles. It stimulates development of the male reproductive tract from the Wolffian (or mesonephric) ducts, which would otherwise degenerate, as occurs in female development. After conversion into dihydrotestosterone (DHT) through action of 5α-reductase in target tissues, it triggers development of the male external genitalia. It also influences brain development and other male-specific

characteristics, although in many instances this is also thought to be indirect, with the testosterone being converted to oestrogens via local action of aromatase (Forger 2001, Gooren & Kruijver 2002, Han & De Vries 2003, Negri-Cesi et al 2004).

- Insulin-like factor 3 (INSL3) is also made by Leydig cells in the fetus. It is thought to have a very specific role in promoting the descent of the testes from a position near the kidneys into the scrotum (Nef & Parada 1999, Zimmermann et al 1999).

High levels of androgens are thought to be responsible for many aspects of male behaviour, including aggression, but they could act at several different times, in different ways and on different processes (Gooren & Kruijver 2002, Swaab 2004). Androgens are likely to act in the embryo to specify sex-specific neuroanatomy and circuitry, which would result in stable behavioural differences. This role for androgens is clearly seen in animal studies, but how much it is true for humans is controversial (Palanza et al 1999, Gooren & Kruijver 2002, Rowe et al 2004, Swaab 2004). There are, of course, few experimental data on administering steroid hormones to human embryos *in utero*. There are studies on a large cohort of children born from women given diethylstilbestrol (DES), a synthetic oestrogen, to help maintain difficult pregnancies (Golden et al 1998, Wilcox et al 1995). Originally it was thought that higher than normal numbers of these children were discordant with respect to their phenotypic sex, gender identity, gender preference and gender role. More recent studies, based on larger numbers of individuals, have suggested that any effects are very weak or non-existent (Titus-Ernstoff et al 2003). On the other hand, with several types of intersex development, for example in congenital adrenal hyperplasia (CAH), where 21-hydroxylase mutations lead to androgen synthesis by the adrenal, this can have a masculinizing effect not only on the external genitalia of girls but also on their behaviour, which tends to be 'tomboyish' (Grumbach et al 2003). However, again, prenatal exposure seems to have weaker effects than might be expected and it is suspected that environmental or social conditioning can have a strong overriding effect. Indeed, this idea led to another dogma, championed by John Money, that effects of hormones on the brain are not fixed in humans until a few years after birth, and can be overridden by environmental conditioning. In other words if surgery and/or appropriate hormone treatments are given sufficiently early after birth, the child will grow up according to the sex to which it was assigned and raised. This theory was brought into dispute with several well known and probably many less well-known cases where gender identity did not match the reassigned sex (Hrabovszky & Hutson 2002, Conte & Grumbach 2003, Grumbach et al 2003). Clearly, much is not known about how and when androgens act on behaviour during development, or on their relative importance in different species.

Steroid hormones, including androgens, can also act in the adult animal to modulate neuronal activity, stem cell self-renewal and differentiation in the CNS, and, in some species, to promote seasonal changes in neuroanatomy. Many studies have looked at the role of testosterone in social behaviours, often with the notion that high levels are likely to lead to aggression (for example, O'Connor et al 2004). However, many results have been far from clear. Indeed some studies suggest that high testosterone levels are a consequence rather than a cause of aggressive behaviour (Ross et al 2004). On the other hand, much of the confusion probably relates to the fact that data has been obtained in many different species with many different assays, and the exact role testosterone plays in aggression will depend on the animal under study.

Finally, through their action on the rest of the body, androgens also lead to the development of many of the morphological characteristics required for male-associated behaviours and, which may indeed help to promote them. For example, the larger size and strength of males, means they are less likely to be deterred from behaving aggressively.

SRY, sex determination and the origin and differentiation of Leydig cells

Given that Leydig cells in the testis are an important and male-specific source of androgens, it is relevant to understand how they develop with respect to the process of sex determination and the origin of other cells types in the gonads. SRY acts at about 11 days post coitum (dpc) in the mouse or at 6 weeks in humans (Lovell-Badge et al 2002). Evidence from a variety of sources reveals that it functions as a trigger specifically within cells of the supporting cell lineage to initiate their differentiation into Sertoli cells, characteristic of the testis, rather than follicle cells found in the ovary (Sekido et al 2004). Once these cells begin to differentiate, they trigger the other cell lineages to follow a male pathway. These include two other bipotential cell types already present within the gonad: the germ cells arrest in mitosis and become prospermatogonia and steroidogenic cell precursors become Leydig cells. The latter begin to differentiate and express steroidogenic enzymes by about 12.5 dpc in the mouse. Other cell types, notably peritubular myoid and vascular endothelial cells, migrate into the early testis only subsequent to SRY action (Brennan & Capel 2004). It has become increasingly clear that there is considerable interdependence between the various cell types of the gonad, such that defects in cells that differentiate later can affect the maintenance of earlier cell types, leading to secondary gonadal dysgenesis and sex reversal.

The various cell types organize into the typical architecture of the testis, with the epithelial testis cords (which mature into the seminiferous tubules) comprised of germ cells surrounded by Sertoli cells and then a layer of myoid cells. Leydig and endothelial cells are within the interstitium, although the latter also give rise to a male-specific artery under the coelomic surface of the testis (Brennan et al 2002, Brennan & Capel 2004). In the genital ridge and the ovary, there is only a capillary bed with little directional blood flow. On the other hand, the male-specific vasculature, which is formed over a 24 hour period, directs blood flow from the coelomic vessel through the testis towards the underlying mesonephros. It seems likely that this serves to rapidly and efficiently export factors, notably testosterone, to the rest of the embryo.

Early gonadal development depends on a number of transcription factors, including SF1 and WT1. These same factors also directly activate gene expression at several subsequent stages or continuously within the supporting cell lineage. Homozygous null mutations in the genes encoding them lead to degeneration of the genital ridge (through apoptosis) at about 11.5 dpc. However, certain types of hypomorphic mutation can lead to XY female sex reversal suggesting that these genes are more critical for testicular than ovarian development (Barbaux et al 1997, Achermann et al 1999, Hossain & Saunders 2001). Indeed they are required for activating genes in the male pathway, which begins with *Sry*. SRY action leads to a rapid up-regulation of the related gene *Sox9*, which has been shown by mutation studies to be critical for testis development (Foster et al 1994, Chaboissier et al 2004). SOX9 then probably directly activates the expression of several downstream effectors of Sertoli cell differentiation and male development, although *Amh* and *Vanin1*, which encodes a membrane-associated protein, are the only proven target genes (De Santa Barbara et al 1998, Arango et al 1999, Wilson et al 2005). SOX9 expression also leads to a rapid down-regulation of several genes, including *Sry* itself, which is only on for a few hours in each Sertoli cell precursor (Chaboissier et al 2004, Sekido et al 2004). In contrast, once it reaches a critical threshold, SOX9 remains expressed at high levels in Sertoli cells throughout life, and this may depend at least in part on autoregulation. Activation of SOX9 in supporting cells in the absence of SRY can lead to XX male development, implying that *Sox9* is the only critical gene downstream of SRY (Huang et al 1999, Bishop et al 2000, Vidal et al 2001). Alternatively, other cases of XX male development imply that there must be genes that act as ovary determining or 'anti-testis' factors (Lovell-Badge et al 2002, Yao 2005). In normal female development these genes may promote follicle cell differentiation directly, but they must also serve to repress genes required for male development downstream of SRY. There are several candidates for such ovary determining genes, notably *Dax1*, *Wnt4*, follistatin and *FoxL2*, all of which are expressed in the indifferent genital ridge and stay on in the ovary, while they are repressed as Sertoli cells differentiate in the testis (Swain et al 1998, Yao 2005). However, loss of function mutations in each of these only gives at best partial forms of XX male sex reversal (Jeays-

Ward et al 2003, Meeks et al 2003, Schmidt et al 2004, Uda et al 2004, Yao et al 2004).

Once Sertoli cells form they must actively trigger or permit Leydig cell differentiation. SF1 is thought to be critical for the transcriptional regulation of many genes in Leydig cells, including those of the steroidogenic pathway and *Insl3*, but these cells are likely to have a different origin to those of the supporting cell lineage (Jeyasuria et al 2004, Morohashi 1997, Karl & Capel 1998). Several signalling molecules and pathways have also been implicated in Leydig cell differentiation. Desert hedgehog (DHH) is expressed by Sertoli cells as soon as they begin to differentiate. *Dhh* mutants show a reduction or absence of Leydig cells, and later germ cells defects, although other testicular cell types appear normal (Bitgood et al 1996, Yao et al 2002). Platelet-derived growth factor receptor-α (*Pdgfr-α*) mutants are complex, showing a reduction in general proliferation, cell migration, cord formation, but also of Leydig cell differentiation. In *Wnt4* mutants, steroidogenic cells are found in the XX gonad. These behave like Leydig cells, producing testosterone, which in turn leads to signs of masculinization. However, this is due to the abnormal migration of cells from the anterior mesonephros that are normally destined for the adrenal. In the absence of *Wnt4* these cells are permitted to enter the ovary along with endothelial cells, such that the latter give rise to the typical male pattern of vasculature. However, there are no signs of Sertoli cell differentiation or cord formation (Jeays-Ward et al 2003).

The number of Leydig cells that differentiate and their output depends in part on Sertoli cell number: for example, the presence of ovotestes often results in intersex development. Moreover, the efficiency with which testosterone is exported from the testis will probably depend on the correct establishment of the vasculature, although the latter may have additional roles helping to specify the organization of the testis into cords.

Around or shortly after birth, fetal Leydig cells are replaced by adult Leydig cells. The origin of these cells is unknown, but they are assumed to arise from one of several interstitial cell types that appear undifferentiated in the fetal testis (Ge et al 2005). One interesting possibility is that they arise from a resident stem cell population. The output of the adult Leydig cells is regulated by a number of external influences, both from adjacent cell types within the testis, but also via the pituitary/hypothalamic axis. These can therefore all indirectly modify an animal's behaviour.

Two unconventional mammalian species with 'masculinized' females: Hyenas and moles

While the majority of mammals show sex determination and differentiation mechanisms similar to those in mice and humans, there are a number of exceptions. It is worth describing two that are relevant to the topic of aggression.

Males of the garden mole, *Talpa occidentalis* have normal testes, however females always develop ovotestes, with a functional, but often minor ovarian portion, and a testicular portion (Barrionuevo et al 2004). The latter possesses Leydig cells, peritubular myoid cells and a male type vasculature, but lacks germ cells and shows no markers typical of Sertoli cells. The females have highly elevated testosterone levels, but internal and external genitalia are not masculinized. This suggests that the male hormone controls some aspect of behaviour required for living a largely solitary life underground. Aggression seems to be a distinct possibility. Interestingly, during the mating season, the testicular portion of the ovotestes regresses, and it is thought to be only during this time that the females allow males to approach. Unfortunately, their lifestyle makes it difficult to study mole behaviour and they cannot be bred or even kept in captivity. The molecular cause of the ovotestis development is not understood, although the females lack *Sry* and it does not appear to depend on ectopic activation of *Sox9*.

Females of the spotted hyena, *Crocuta crocuta*, have masculinized external genitalia and are very aggressive! Maternal androstenedione, which has an ovarian origin, is converted by the placenta to testosterone from early/mid gestation stages up to birth and this plays a critical part in the masculinization. In other words, this is an example of a strong effect of exogenous steroids on fetal development and subsequent behaviour (Drea et al 2002).

This masculinization causes problems with mating and with birth (both are mechanically awkward!). There is a high rate of neonatal mortality in firstborn litters due to anoxia. Oral administration of anti-androgens (finasteride and flutamide) to pregnant hyenas leads to feminization of the 'phallus' in female offspring (it becomes shorter, thicker and more elastic), and all their firstborn survive. However, male offspring have a penis that is too short and the wrong shape for intromission — they are essentially infertile. This therefore constitutes an interesting example of morphological co-evolution, where both sexes depend on high levels of androgens, even though this has significant effects on mortality (Drea et al 2002).

The contribution of other Y and X chromosome genes to differences between the sexes

The 'dogma' described above suggests that SRY action leads to Sertoli cell differentiation. These cells then trigger the differentiation of other testicular cell types, including Leydig cells, which in turn produce high levels of testosterone (modified to DHT or oestrogen at some target tissues) leading to all other male characteristics in terms of anatomy, physiology and behaviour. But does the dogma account for all differences between males and females?

Other factors produced by the gonads could play a role. For example, Sertoli cells produce AMH, which apart from its eponymous role to eliminate the Mullerian ducts, is also known to have a back-up function to eliminate any meiotic germ cells from the fetal testis (Josso et al 1998). Perhaps it could have other, more systemic effects. Leydig cells also produce *Insl3*, required for testis descent; does this do anything else?

It is also possible, however, that in addition to *Sry*, other sex-linked genes, on either the X or Y chromosome, could exert direct effects modifying any aspect of male or female anatomy, physiology and behaviour. In theory, these genes could act synergistically with gonadal secretions, in which case their effects will follow most phenotypic sex differences. For example, a gene from the Y could confer some male behaviour, but only in the presence of high testosterone and androgen receptors, or high oestrogen levels and oestrogen receptors. Alternatively, the genes could act independently, such that any phenotypes they produce will be independent of hormone-induced phenotypic sex differences.

Y-linked genes

Sex chromosomal genes with direct effects on sex differences could map to the non-recombining region of the Y and either be unique to the Y chromosome or have an X-linked homologue, but where the Y gene has acquired a new function (due to regulation or protein structure). There is no reason why genes with roles in the gonad, such as spermatogenesis, could not have other functions, in e.g. the brain. This would include *Sry*, and there is some evidence that this is expressed in parts of the CNS (Mayer et al 2000, Reisert et al 2002). Such roles may be species specific, indeed they may be under high selective pressure and evolve rapidly, especially if they confer a mating advantage. A number of sex-specific phenotypes have already been attributed to Y chromosome genes.

In humans, *AMELY* (encodes amelogenin) gives a male-specific pattern to tooth enamel (Fincham et al 1991). There is also good evidence for a Y-linked gene involved in stature (Kirsch et al 2004). (This is one reason why XY individuals with androgen insensitivity syndrome—testicular feminization—tend to be taller on average than XX women). There is also some evidence for a gene involved in high blood pressure (Rodriguez et al 2005). In rodents there is good evidence for Y-linked genes having effects on a range of behaviours and on CNS development (De Vries et al 2002, Arnold & Burgoyne 2004). These include:

- Control of aggressive and copulatory behaviours
- Hippocampal size asymmetry

- Mossy fibre distribution
- Open-field activity
- Apomorphine-induced stereotyped behaviours, and
- Entrainment of circadian rhythms.

With respect to Y chromosome effects on aggressive behaviours in mice there have been many studies, but often with conflicting results. A selected sample of publications reveals this (N.B. PAR refers to the pseudoautosomal region of the Y and X chromosomes, which contain homologous sequences involved in pairing and exchange during male meiosis. NPAR refers to the non-PAR or unique portion of the Y):

- Robertoux et al (1994) showed that in comparisons of males of the NZB and CBA strains, the initiation of attack behaviour is not due to the Y^{NPAR}, but the Y^{PAR} and autosomes have additive effects.
- Guillot & Chapouthier (1998), also failed to find Y^{NPAR} effects. Males of more attacking strains have lower olfactory thresholds, which could make olfactory discrimination of opponents easier and identification of strangers more efficient.
- Monahan & Maxson (1998) looked at mice congenic for DBA and C57BL10 Y chromosomes on a DBA background. They found several Y^{NPAR} effects on likelihood of offense, affecting (i) motivating stimuli (testosterone-dependent pheromones and non-testosterone-dependent (urine) odour-types), (ii) olfactory perception of motivating stimuli and (iii) the motivational mechanism of offense.
- Sluyter et al (1999) found Y^{NPAR} influences on defensive burying (associated with males showing high aggression (short attack latency). The effects were small compared to non-Y effects and females from same selected (SAL versus LAL) strains tended to show the same behavioural profile as males.

X-linked genes

Most X-linked genes are subject to X-inactivation, ensuring that gene dosage is the same between XX and XY individuals. However, several escape inactivation, either generally or in a tissue-specific manner. This is especially true in humans, where perhaps as many as 25% do so (Carrel & Willard 2005). This will give different gene doses between XX and XY individuals. There are a number of examples of behavioural and phenotypic traits, although the genes responsible have yet to be correlated with phenotypes and vice versa.

- In marsupials, the pouch and scrotum develop according to X dosage (XX and X, respectively) and are independent of gonadal sex, which depends on the Y chromosome, where there is an *Sry* homologue. XO animals develop ovaries and a scrotum, while XXY animals have testes and a pouch (Santucciu et al 2003).

- In mice, the density of vasopressin fibres in the lateral septum depends on X dosage and is independent of sex (XX females = XX*Sry* males > XY males = XY females) (De Vries et al 2002).
- In mice, 6 out of 8 X-linked genes (with Y-homologues), e.g. *Usp9x*, were all expressed at higher levels in XX brains compared with XY brains (Xu et al 2002).
- Effects on fear reactivity in XO mice are due to haplo-insufficiency of a non-PAR X gene (Isles et al 2004).

Imprinted X-linked genes

Genes on the maternal and paternal X chromosomes may behave differently due to the phenomenon of genomic imprinting. This is known for the mouse *Xist* gene, which is preferentially active on the paternal X in extraembryonic tissues. Again there are a number of examples with phenotypic consequences, although the genes involved are still to be identified.

- In mice, there is a difference in developmental rate between male and female postimplantation embryos that is due to a retarding effect of the paternal X chromosome, such that $XY = X^mO > X^mX^p > X^pO$ (Thornhill & Burgoyne 1993).
- With human Turner patients, X^mO and X^pO individuals tend to show behavioural differences and memory functions (Skuse et al 1997).

These effects would also be expected in cases of skewed X-inactivation in XX or XXY individuals. Skewing can occur by chance or due to a deleterious mutation on one X.

Conclusions

Accumulating evidence suggests that X and Y linked genes may contribute to differences between the sexes in anatomy, physiology and behaviour. It is not clear how important these are or if they would compound decisions when treating cases of sex reversal and intersex conditions (Conte & Grumbach 2003). Evidence that there are genes involved in aggression mapping to the sex chromosomes is often contentious. Certainly, proposals that XYY men are more aggressive have been refuted. However, it is possible that higher than normal levels of SRY could speed up testis development leading to a greater recruitment of

Leydig cells, and therefore to higher levels of testosterone. Similarly, other Y-linked genes, some of which are expressed ubiquitously, could affect brain development. Y chromosomal genes affecting stature or ability to interact with others could also have indirect effects on behaviour. It also seems that X-linked genes could directly impact on aggressive behaviours. The gene encoding the androgen receptor maps to the X chromosome, so allelic variation at this locus could have direct effects. It is also clear that many cases of mental retardation are due to mutations in X-linked genes, so it would not be surprising if allelic differences in such genes had more or less subtle effects on behaviour. Perhaps the strongest conclusion to be reached is that much more research is needed on this topic. If the effects being measured are subtle then care must be taken to reduce compounding effects of genetic background; moreover it is important to establish model systems where sex chromosome influences can be looked at independently of gonadal sex (De Vries et al 2002).

Acknowledgements

I am grateful to Paul Burgoyne, Vasso Episkopou and members of my laboratory for valuable discussions.

References

Achermann JC, Ito M, Hindmarsh PC, Jameson JL 1999 A mutation in the gene encoding steroidogenic factor-1 causes XY sex reversal and adrenal failure in humans. Nat Genet 22:125–126

Arango NA, Lovell-Badge R, Behringer RR 1999 Targeted mutagenesis of the endogenous mouse Mis gene promoter: in vivo definition of genetic pathways of vertebrate sexual development. Cell 99:409–419

Arnold AP, Burgoyne PS 2004 Are XX and XY brain cells intrinsically different? Trends Endocrinol Metab 15:6–11

Barbaux S, Niaudet P, Gubler MC et al 1997 Donor splice-site mutations in WT1 are responsible for Frasier syndrome. Nat Genet 17:467–470

Barrionuevo FJ, Zurita F, Burgos M, Jimenez R 2004 Testis-like development of gonads in female moles. New insights on mammalian gonad organogenesis. Dev Biol 268:39–52

Bishop CE, Whitworth DJ, Qin Y et al 2000 A transgenic insertion upstream of Sox9 is associated with dominant XX sex reversal in the mouse. Nat Genet 26:490–494

Bitgood MJ, Shen L, McMahon AP 1996 Sertoli cell signaling by Desert hedgehog regulates the male germline. Curr Biol 6:298–304

Brennan J, Capel B 2004 One tissue, two fates: molecular genetic events that underlie testis versus ovary development. Nat Rev Genet 5:509–521

Brennan J, Karl J, Capel B 2002 Divergent vascular mechanisms downstream of Sry establish the arterial system in the XY gonad. Dev Biol 244:418–428

Carrel L, Willard HF 2005 X-inactivation profile reveals extensive variability in X-linked gene expression in females. Nature 453:400–404

Chaboissier MC, Kobayashi A, Vidal VI et al 2004 Functional analysis of Sox8 and Sox9 during sex determination in the mouse. Development 131:1891–1901

Conte FA, Grumbach MM 2003 Diagnosis and management of ambiguous external genitalia. Endocrinologist 113:260–268
De Santa Barbara P, Bonneaud N, Boizet B et al 1998 Direct interaction of SRY-related protein SOX9 and steroidogenic factor 1 regulates transcription of the human anti-Mullerian hormone gene. Mol Cell Biol 18:6653–6665
De Vries GJ, Rissman EF, Simerly RB et al 2002 A model system for study of sex chromosome effects on sexually dimorphic neural and behavioral traits. J Neurosci 22:9005–9014
Drea CM, Place NJ, Weldele ML, Coscia EM, Licht P, Glickman SE 2002 Exposure to naturally circulating androgens during foetal life incurs direct reproductive costs in female spotted hyenas, but is prerequisite for male mating. Proc R Soc Lond B Biol Sci 269:1981–1987
Fincham AG, Bessem CC, Lau EC et al 1991 Human developing enamel proteins exhibit a sex-linked dimorphism. Calcif Tissue Int 48:288–290
Forger NG 2001 The development of sex differences in the nervous system. In: Blass EM (ed) Handbook of behavioural neurobiology. Vol 12. Plenum, New York, p 153–208
Foster JW, Dominguez-Steglich MA, Guioli S et al 1994 Campomelic dysplasia and autosomal sex reversal caused by mutations in an SRY-related gene. Nature 372:525–530
Ge RS, Dong Q, Sottas CM, Chen H, Zirkin BR, Hardy MP 2005 Gene expression in rat leydig cells during development from the progenitor to adult stage: a cluster analysis. Biol Reprod 16, in press
Golden RJ, Noller KL, Titus-Ernstoff L et al 1998 Environmental endocrine modulators and human health: an assessment of the biological evidence. Crit Rev Toxicol 28:109–227
Gooren LJ, Kruijver FP 2002 Androgens and male behavior. Mol Cell Endocrinol 198:31–40
Grumbach MM, Hughes IA, Conte FA 2003 Disorders of sex differentiation. In: Larsen PR, Kronenberg HM, Melmed S, Polonsky KS (eds) Williams textbook of endocrinology, 10th edn. WB Saunders, Philadelphia, p 842–1002
Guillot PV, Chapouthier G 1998 Intermale aggression, GAD activity in the olfactory bulbs and Y chromosome effect in seven inbred mouse strains. Behav Brain Res 90:203–206
Han TM, De Vries GJ 2003 Organizational effects of testosterone, estradiol, and dihydrotestosterone on vasopressin mRNA expression in the bed nucleus of the stria terminalis. J Neurobiol 54:502–510
Hossain A, Saunders GF 2001 The human sex-determining gene SRY is a direct target of WT1. J Biol Chem 276:16817–16823
Hrabovszky Z, Hutson JM 2002 Androgen imprinting of the brain in animal models and humans with intersex disorders: review and recommendations. J Urol 168:2142–2148
Huang B, Wang S, Ning Y, Lamb AN, Bartley J 1999 Autosomal XX sex reversal caused by duplication of SOX9. Am J Med Genet 87:349–353
Isles AR, Davies W, Burrmann D, Burgoyne PS, Wilkinson LS 2004 Effects on fear reactivity in XO mice are due to haploinsufficiency of a non-PAR X gene: implications for emotional function in Turner's syndrome. Hum Mol Genet 13:1849–1855
Jeays-Ward K, Hoyle C, Brennan J et al 2003 Endothelial and steroidogenic cell migration are regulated by WNT4 in the developing mammalian gonad. Development 130: 3663–3670
Jeyasuria P, Ikeda Y, Jamin SP et al 2004 Cell-specific knockout of steroidogenic factor 1 reveals its essential roles in gonadal function. Mol Endocrinol 18:1610–1619
Josso N, Picard JY 1986 Anti-Mullerian hormone. Physiol Rev 66:1038–1090
Josso N, Racine C, di Clemente N, Rey R, Xavier F 1998 The role of anti-Mullerian hormone in gonadal development. Mol Cell Endocrinol 145:3–7
Jost A 1953 Problems in fetal endocrinology; the gonadal and hypophyseal hormones. Rec Prog Horm Res 8:379–418
Karl J, Capel B 1998 Sertoli cells of the mouse testis originate from the coelomic epithelium. Dev Biol 203:323–333

Kirsch S, Weiss B, Zumbach K, Rappold G 2004 Molecular and evolutionary analysis of the growth-controlling region on the human Y chromosome. Hum Genet 114:173–181

Koopman P, Gubbay J, Vivian N, Goodfellow P, Lovell-Badge R 1991 Male development of chromosomally female mice transgenic for Sry. Nature 351:117–121

Lovell-Badge R, Canning C, Sekido R 2002 Sex-determining genes in mice: building pathways. In: The genetics and biology of sex determination. Wiley, Chichester (Novartis Found Symp 244) p 4–22

Luo X, Ikeda Y, Parker KL 1994 A cell-specific nuclear receptor is essential for adrenal and gonadal development and sexual differentiation. Cell 77:481–490

Mayer A, Mosler G, Just W, Pilgrim C, Reisert I 2000 Developmental profile of Sry transcripts in mouse brain. Neurogenetics 3:25–30

Meeks JJ, Weiss J, Jameson JL 2003 Dax1 is required for testis determination. Nat Genet 34:32–33

Monahan EJ, Maxson SC 1998 Y chromosome, urinary chemosignals, and an agonistic behavior (offense) of mice. Physiol Behav 64:123–132

Morohashi K 1997 The ontogenesis of the steroidogenic tissues. Genes Cells 2:95–106

Munsterberg A, Lovell-Badge R 1991 Expression of the mouse anti-mullerian hormone gene suggests a role in both male and female sexual differentiation. Development 113:613–624

Nef S, Parada LF 1999 Cryptorchidism in mice mutant for Insl3. Nat Genet 22:295–299

Negri-Cesi P, Colciago A, Celotti F, Motta M 2004 Sexual differentiation of the brain: role of testosterone and its active metabolites. J Endocrinol Invest 27(6 suppl):120–127

O'Connor DB, Archer J, Wu FC 2004 Effects of testosterone on mood, aggression, and sexual behavior in young men: a double-blind, placebo-controlled, cross-over study. J Clin Endocrinol Metab 89:2837–2845

Palanza P, Parmigiani S, Liu H, vom Saal FS 1999 Prenatal exposure to low doses of the estrogenic chemicals diethylstilbestrol and o,p′-DDT alters aggressive behavior of male and female house mice. Pharmacol Biochem Behav 64:665–672

Reisert I, Karolczak M, Beyer C, Just W, Maxson SC, Ehret G 2002 Sry does not fully sex-reverse female into male behavior towards pups. Behav Genet 32:103–111

Rodriguez S, Chen XH, Miller GJ, Day IN 2005 Non-recombining chromosome Y haplogroups and centromeric HindIII RFLP in relation to blood pressure in 2,743 middle-aged Caucasian men from the UK. Hum Genet 116:311–318

Ross CN, French JA, Patera KJ 2004 Intensity of aggressive interactions modulates testosterone in male marmosets. Physiol Behav 83:437–445

Roubertoux PL, Carlier M, Degrelle H, Haas-Dupertuis MC, Phillips J, Moutier R 1994 Co-segregation of intermale aggression with the pseudoautosomal region of the Y chromosome in mice. Genetics 136:225–230

Rowe R, Maughan B, Worthman CM, Costello EJ, Angold A 2004 Testosterone, antisocial behavior, and social dominance in boys: pubertal development and biosocial interaction. Biol Psychiatry 55:546–552

Santucciu C, Grutzner F, Carvalho-Silva DR, Graves JA 2003 Isolation of chromosomal regions controlling intersex development in a marsupial. Cytogenet Genome Res 101:224–228

Schmidt D, Ovitt CE, Anlag K et al 2004 The murine winged-helix transcription factor Foxl2 is required for granulosa cell differentiation and ovary maintenance. Development 131: 933–942

Sekido R, Bar I, Narvaez V, Penny G, Lovell-Badge R 2004 SOX9 is up-regulated by the transient expression of SRY specifically in Sertoli cell precursors. Dev Biol 274:271–279

Skuse DH, James RS, Bishop DV et al 1997 Evidence from Turner's syndrome of an imprinted X-linked locus affecting cognitive function. Nature 387:705–708

Sluyter F, Korte SM, Van Baal GC, De Ruiter AJ, Van Oortmerssen GA 1999 Y chromosomal and sex effects on the behavioral stress response in the defensive burying test in wild house mice. Physiol Behav 67:579–585
Swaab DF 2004 Sexual differentiation of the human brain: relevance for gender identity, transsexualism and sexual orientation. Gynecol Endocrinol 19:301–312
Swain A, Narvaez V, Burgoyne P, Camerino G, Lovell-Badge R 1998 Dax1 antagonizes Sry action in mammalian sex determination. Nature 391:761–767
Thornhill AR, Burgoyne PS 1993 A paternally imprinted X chromosome retards the development of the early mouse embryo. Development 118:171–174
Titus-Ernstoff L, Perez K, Hatch EE et al 2003 Psychosexual characteristics of men and women exposed prenatally to diethylstilbestrol. Epidemiology 14:155–160
Uda M, Ottolenghi C, Crisponi L et al 2004 Foxl2 disruption causes mouse ovarian failure by pervasive blockage of follicle development. Hum Mol Genet 13:1171–1181
Vidal VP, Chaboissier MC, de Rooij DG, Schedl A 2001 Sox9 induces testis development in XX transgenic mice. Nat Genet 28:216–217
Wilcox AJ, Baird DD, Weinberg CR, Hornsby PP, Herbst AL 1995 Fertility in men exposed prenatally to diethylstilbestrol. N Engl J Med 332:1411–1416
Wilson MJ, Jeyasuria P, Parker KL, Koopman P 2005 The transcription factors steroidogenic factor-1 and SOX9 regulate expression of Vanin-1 during mouse testis development. J Biol Chem 280:5917–5923
Xu J, Burgoyne PS, Arnold AP 2002 Sex differences in sex chromosome gene expression in mouse brain. Hum Mol Genet 11:1409–1419
Yao HH 2005 The pathway to femaleness: current knowledge on embryonic development of the ovary. Mol Cell Endocrinol 230:87–93
Yao HH, Whoriskey W, Capel B 2002 Desert Hedgehog/Patched 1 signaling specifies fetal Leydig cell fate in testis organogenesis. Genes Dev 16:1433–1440
Yao HH, Matzuk MM, Jorgez CJ et al 2004 Follistatin operates downstream of Wnt4 in mammalian ovary organogenesis. Dev Dyn 230:210–215
Zimmermann S, Steding G, Emmen JM et al 1999 Targeted disruption of the Insl3 gene causes bilateral cryptorchidism. Mol Endocrinol 13:681–191

DISCUSSION

Pfaff: What do we know about the cascade of events that follows on from *Sry* expression, and what do we not know? What you have just told us is that *Sox9* is a crucial and sufficient second step. What happens after that?

Lovell-Badge: We know *Sox9* is involved in turning on a number of genes. The one that has been best studied is *Amh*. We know that *Sox9* up-regulation is essential for expression of that gene. There are others such as hedgehog that come on at the same time, and it is likely they are direct targets of *Sox9*. We know a whole pile of transcription factors that are involved in allowing a gonad to develop in either sex. These are things like the Wilms tumour gene, *Sf1*, *Gata4* and *Lhx9*: these are all required to make a gonad. There are certain types of mutations in these genes that can lead to sex reversal to give XY females, suggesting that these genes have a slightly more important role in male development than female development. In fact, they are probably involved in both. All these genes work generally in the pathway. Usually as *Sry* comes on it turns *Sox9* up and there must be regulatory loops that maintain the high level of *Sox9*, because *Sry* goes off. It is possible that there is a direct autoregulatory loop by which *Sox9* maintains itself.

Pfaff: Is the promoter element well established?

Lovell-Badge: We have demonstrated cord development using a *Sox9* regulatory element to drive CFP. We think this is the critical element, but it has taken about four years to find this, because the regulatory region of *Sox9* is spread over at least a megabase. We think we have that critical region. In this region we find potential binding sites for SRY, SF1 and perhaps GATA4. These factors may all have to be present to up-regulate *Sox9*. It has been suggested that there is a regulatory loop: SF1 is contributing to turning SOX9 on, and once SOX9 gets above a threshold it feeds back and up-regulates SF1. There are all these possibilities for loops to maintain expression. What we are missing are the genes that are making ovaries; that work against the male pathway. There have been some candidates, but none of them have worked out in mouse experiments. There is a candidate gene found in cases of sex reversal in goats where XX males are produced. Mutating this gene in mice has no effect.

Dulac: What is the advantage of *Sry* mutant rescue by using a *Sry* transgene?

Lovell-Badge: When one makes a transgenic mouse, there is a possibility of two things happening. One is that you might not get the same expression from the transgene as you would from the endogenous gene, so this is a control for that.

Dulac: Do you have a transgene that expresses only in the testis but not in the brain?

Lovell-Badge: We can do (and have done) this. We take the mouse genomic fragment (sequences expressing the mouse gene plus the regulatory region; this is what most transgenics have been made with). If we use the human gene it doesn't give sex reversal, but if we take the human coding region and put it under control of the mouse regulatory sequence it does work, so we can compare how mouse and human *Sry* work. *Sry* is one of the most rapidly evolving genes. You can't even line up the two protein or DNA sequences for human and mouse *Sry*, except for just one part of the protein, the DNA binding region. Even this is quite divergent. The experiment in which we can successfully swap the human into mouse tells us that for sex determination the function is conserved, but the proteins are very different. One theory for explaining this difference is that *Sry* might be doing something else that accounts for its rapid rate of evolution.

Dulac: Is there any phenotype in mice that lack *Sry* expression in the brain?

Lovell-Badge: There are several ways we can make mice without *Sry* in the brain. One is to use a different promoter. Rather than using the *Sry* promoter we can use the *Dax1* promoter to drive *Sry*. This will initiate male development perfectly normally. We haven't looked at whether there are differences. There are also a number of ways we can make mice male without SRY being present at all. The best way to do this is to activate *Sox9* without SRY being present. There is a mutation that Colin Bishop has worked on in Baylor called *Odd sex* (Bishop et al 2000). It is an insertion that leads to an up-regulation of *Sox9* in the gonad in the absence of SRY. It gives XX males.

Nelson: You can also get *Sox9* knockouts. They have no gonads.

Lovell-Badge: They all develop as XY females. The problem is that they don't live. This has to be done in such a way that the knockout is restricted just to the developing gonad, and no one has been able to do this yet.

Skuse: I read a paper a while ago presaging the death of the Y chromosome altogether (Aitken & Marshall Graves 2002). I appreciate this point of view has since been questioned (Hawley 2003). On the other hand, isn't there a vole in which the Y doesn't play any role in sex determination (Marshall Graves 2002)? Does this affect their aggressive behaviour?

Lovell-Badge: I don't know. This is the mole vole. There are three closely related species, one of which is normal XX/XY with *Sry* on the Y chromosome, as in the mouse. The other two have both lost the Y chromosome. I wish we could get material to look at this. I bet this is something like an activating mutation of *Sox9*. They will probably now evolve another sex chromosome if we wait long enough!

Craig: You raised the question about escaping from X inactivation. I think not enough emphasis is placed on this in the sense that recent accounts suggest that some 20% of the genes in some tissues are going to escape from X inactivation. Most of the work on this has been done in humans; what do we know about this in mice?

Lovell-Badge: There are relatively few genes which escape X inactivation in the mouse. This is a big difference between mouse and humans.

Craig: There are genes on the X chromosome, which may cause interesting effects by skewed inactivation. We have been working on monozygous female twins. If there are genes on the X chromosome which are influencing behaviours in some kind of way, you would predict that they are more likely to be discordant than male monozygous twins. The answer is that if you look at large number of twins, there are two or three behaviours that are much more discordant in the female monozygous twins, one of which has to do with peer relations. This refers to work done with Robert Plomin.

Pfaff: Could you tell us more?

Craig: The twins are studies at 2, 3 and 4 so they are still young. We looked at nine behaviours and peer relations at age 4, verbal ability at age 2 and pro-social behaviour at age 2 were the three features that we found to be significantly discordant.

Suomi: There is some nice work looking at species differences among voles in parental behaviour and affiliative behaviour patterns. Has anyone looked at some of these early differentiating genes to see the possible magnitude of differences in these species? In some cases males are very similar to females in a wide range of behaviours, while in other closely related species they are completely different.

Lovell-Badge: As far as I know, no one has.

Suomi: Sry is one of the most rapidly evolving genes. Has anyone looked at differences between humans and other primates, or within the primate order?

Lovell-Badge: They have looked at the rate of evolution of the sequence. It is close enough that you can line it up in primates, and you can work out rates of change. In that study, if you ignore the DNA-binding domain which is reasonably conserved, the rates of change are so rapid and of a particular type that this suggests there was selection for change. If there is selection for change, it implies this is having some function. We did a mouse experiment where we put the human protein coding region under the control of mouse regulatory sequences, this is sufficient to give you sex determination. All this says is that you need an HMG box DNA-binding domain of the right sort to induce male development. In one case there are studies saying there is probably selection for function in the other parts of the protein, and these experiments saying that the only bit needed is the DNA binding protein for a testis. This suggests it is doing something else, which is being selected. It could be in spermatogenesis or mate preference.

Pfaff: Talking about the evolution of *Sry* and the shrinking Y chromosome, would you summarize and criticize the work from David Page's lab about the evidence that the Y chromosome has evolved in four discrete steps?

Lovell-Badge: I am not sure I can criticise it that much. These are very difficult studies to do. This was mostly done by comparing mouse and human, with a bit of data from other species. The mouse Y chromosome is still not fully sequenced. With regard to the disappearing Y hypothesis, which Jennie Graves was one of the big proponents of, I have never particularly liked it. You could turn this argument the other way round and ask how many mammalian species have lost their Y chromosome in 200 million years? We know of perhaps three. The chances of us losing it are equally low. David Page's work suggested that one reason we may not be in danger of losing the Y is because we have large inverted repeats which allow for correction through intrachromosomal recombination or repair mechanisms. We are therefore less likely to lose gene function through random mutation than if we had no possibility of pairing and correction.

C Blanchard: I am intrigued by your suggestion that this information could be useful in predicting the effects of surgical reassignment of people with intersex conditions. David, also known as John/Joan, the famous child who was reassigned from male to female after a disastrous circumcision accident, recently committed suicide. This makes it clear that these kinds of sex reassignment operations are fraught with all kinds of horrible possibilities. Given that there are perhaps 100 000 people who have had sex reassignment surgery in the world, is anyone going back to study the relevant gene patterns to see whether people who really don't go along with this change have differences than those who do?

Lovell-Badge: There is so little follow up of all these individuals that it is hard to get a feeling for what is likely to be a problem. When do you fix things like gender role, preference and identity? John Money used to think that there were a couple of years where everything was very labile, and that they would grow up how you raised them. It is now clear that this is not true. There is quite a lot of evidence to say that some intrauterine effects of steroid hormones are important.

Nelson: Two recent studies (Wisniewski et al 2001, Migeon et al 2002) looked at individuals who were born with ambiguous genitalia and were matched for genital phenotype; some individuals were reared as males and some as females. Sex assignment varied as to whom the attending physician was, the religious or cultural background of the family, and so on. One study (Wisniewski et al 2001) involved patients with a congenital micropenis; whereas the other study (Migeon et al 2002) was of male pseudohermaphrodites with truly ambiguous genitalia of various aetiologies. Gender identity was consistent with the sex of rearing and in the rare cases where there was dissatisfaction of gender assignment, it happened as often in those reared male as those reared female. The assignment is usually based on the size of phallus. This is usually related to the timing and amount of hormone exposure, which is related to the genes.

Lovell-Badge: There is a big problem with that study. He was doing retrospective follow-up, and he had a big problem finding people. Of the few he did find, only a small proportion agreed to participate in any study. Most respondents didn't want to have anything to do with him because they were disillusioned with the medical institutions. Only those who didn't feel so bad about what happened to them participated.

Keverne: I want to return to imprinting. I think it is true that *Xist* is imprinted in mice but not humans. It is always the paternal *Xist* allele that is expressed, isn't it?

Lovell-Badge: This is in extraembryonic tissues, where in mice it is the paternal *Xist* that is expressed.

Keverne: As you said, the paternal X carries some factor on it that is growth reducing for the placenta. This is counterintuitive, because most of the paternally expressed autosomal alleles have the opposite effect. This is known from chimeras and parthenogenetic embryos. Even in humans, hydatidiform mole, which is paternal, completely takes over the placenta.

Lovell-Badge: You are talking about placental growth, whereas Paul's study addressed not the placenta, but the embryo itself (Thornhill & Burgoyne 1993).

Keverne: I would like to respond to what Steve was saying about the two vole species, the meadow vole and prairie vole. In a recent paper in *Nature* (Lim et al 2004) the authors transferred the monogamous mole's vasopressin receptor (V1aR) into a mouse. There was a claim that this mouse was then more monogamous. But if we look at these data carefully we see what is really changing. First, the expression in the brain changes: it is important that the mice have the upstream vole sequence in order for this to happen. But in terms of behaviour, the primary change is olfactory investigation. They spend much more time doing this. This in turn affects partner preference and the mouse stays with this partner for longer.

Koolhaas: A couple of years ago we published some evidence (Koolhaas et al 1998) that mechanisms involved in sexual differentiation also play a role in the differentiation within the male gender, in terms of aggressive behaviour and vasopressin in the brain. We didn't continue these studies, but given the amount of detailed knowledge you have now about the molecular mechanisms of gonadal development, what advice would you give regarding the question of where the variation is within one of the sexes?

Lovell-Badge: It is a difficult question to answer, but I would be interested to have a look at what you are doing.

Koolhaas: Some people suggested that we should look for polymorphisms in the *Sry* gene.

Lovell-Badge: Within a species *Sry* is not very polymorphic. I doubt that this is involved.

Pfaff: How big is the promoter?

Lovell-Badge: With an 11 kb genomic fragment we have everything. It has not been well characterized. It is rapidly evolving, so between species you would find big differences. One of the reasons people haven't made much progress defining what is upstream of Sry is because you can't do the usual sort of sequence comparisons to identify common elements.

Brodkin: Just to follow-up on the variability within species, Stephen Maxson at the University of Connecticut has studies implicating the non-recombining part of the Y chromosome as affecting variability among mouse strains in intermale aggression. He proposed *Sry* as a candidate gene that might be responsible for this (Maxson 1996).

Lovell-Badge: It is possible that a higher-expressed or faster-activating version of the *Sry* gene would initiate faster development of the testis, which would then lead to Leydig cells differentiating earlier and perhaps putting out more testosterone. We can find some differences between mice which are male because they have an *Sry* transgene compared with males with the endogenous gene. Different transgenes can also have subtly different effects. This is probably related to level of expression.

Brodkin: What about between inbred strains?

Lovell-Badge: There are differences in levels of expression from different types of Y chromosomes. You have to go to quite different types of Y chromosomes to get a detectable difference. But then you are changing many things besides *Sry*.

Pfaff: I have a couple of questions about the non-pseudoautosomal region (PAR). Steve Maxson has a long series of papers that demonstrate that changes in the non-PAR region can be correlated with changes in aggressive behaviour. According to your map, there are only nine genes there.

Lovell-Badge: There are a few more. There are a few more also on the long arm of the Y chromosome that are not talked about very much. These are in multiple copies. There are certainly other genes on the Y chromosome. Most of them are thought to be playing some role in spermatogenesis. There are multiple copy genes that are very complex to work with.

Pfaff: You also quoted some papers where there are negative results. Could you comment on the genetic artefacts that cause negative results in such studies, with respect to site of integration and so forth. Then I'll ask Jaap and others to comment on potential negative results with respect to how the behavioural assays are carried out.

Lovell-Badge: We tried to do some work with Art Arnold. Initially we were doing these studies with outbreds, because it was going to be difficult to generate all the different genotypes we were interested in using inbreds. One of the problems if you have outbred mice (this may approximate more with humans) is that you need a lot more animals in the study to get statistical significance. This is one source of variation. Clearly the way that they were brought up, and shipped across the Atlantic, will affect the results. It was cheaper to send them in groups. If you keep mice caged together as they grow up, you can keep males together and they are fine. In one solo experiment, Paul Burgoyne had mice which were segregating two different Y chromosomes within the same litter. Males caged together within a litter are usually fine, and don't fight as long as they know each other. But in this case with two different Y chromosomes segregating in the population, they fought each other like mad. It is an interesting observation that needs to be followed up. Whether mice are kept together like this or separated early can affect on how they behave in experiments. Paul has now been transferring some things into inbred background, but the problem with this is that inbreds are quite variable. The best thing to use is F_1s, which complicates your breeding programs.

Pfaff: Regarding behavioural methodology and negative results on non-PARs, what about behavioural results in this sort of study?

Koolhaas: There is an enormous variation in the detailed methods used. Some people isolate animals from the wild and then look at behaviour. The group of Roubertoux has published about how the way one measures aggressive behaviour affects the results. I have no way around this, other than a worldwide agreement on common methodology. There is another problem that hasn't yet been addressed, which is the absolute level of aggressive behaviour. I have seen studies where the variation of aggressive behaviour ranges from 2–4% of the total test time, whereas in our wild-type rats it varies from 0–80% of total test time. It makes a difference if effects of genetic manipulation on aggressive behaviour are shown in this very low range or the total range that is functional in nature. It is a big problem.

Lovell-Badge: We tried reducing variation in some of our experiments by removing the gonads and then giving implants, leading to a constant level of androgens. But this has other consequences.

Craig: Does population genetics say something about aggression and the Y chromosome in humans, in terms of the lineages of Y chromosomes seen sweeping across parts of Europe? Such a feature suggests that there is an attribute (possibly aggression related) associated with a particular haplotype on the Y chromosome.

Suomi: One way around this problem of different methodologies might be videotaping the subjects and having a common record.

Koolhaas: Another way is that you should always present the total ethogram. The behaviour that you present in the paper should cover 100% of the test time. This would let people judge from the behaviour profile whether it is really different from what others observe.

Brodkin: If you go back and look at the papers by Roubertoux and Maxson, one found the non-PAR implicated in aggressive behaviours, and the other found the PAR implicated. They recently co-authored a paper where they attributed some of this discrepancy to differences in their behavioural testing methodologies (Maxson et al 2001).

Nelson: I have a philosophical question relating to the dogma that hormones mediate the sex difference in aggression. Isn't it possible that the dogma might be true here? Everyone is proposing that there will be genes expressed in the brain that have a behavioural effect, but after five or six years few or none have been found.

Lovell-Badge: I think there are effects of X and Y genes.

Skuse: There is indeed increasing evidence for direct effects of X and Y-linked genes on brain development (Carruth et al 2002, Dewing et al 2003, Xu et al 2002). I think however that we will eventually find that the role played by those particular genes in neural development, and ultimately in function, is modulated by hormonal factors. In other words, the impact will be neither wholly genetic nor simply hormonal. I strongly suspect that sexually dimorphic expression of X-linked genes is modulated by sex-steroids, particularly in males.

Nelson: There is some proportion of sex differences attributed to hormones that might be as high as 90%, and those genes expressed directly in the brain might have a small effect. If you read Art Arnold's papers, he seems to suggest that there is a much bigger effect of these genes amounting to a complete paradigm shift in the way we think about the establishment of sexual dimorphism in brain and behaviour. Is that warranted?

Robins: Variations on the edges of the curve might be important. These might not be accounted for by the dogma but might be behaviourally (socially) important. Eric Vilain has data in this regard. Only recently have we had the technology to look at tiny molecular differences in developing brain. Vilain has looked at male–female differences in mouse fetuses prior to the time of sex determination. He has found several genes that are differentially expressed.

Dulac: The differences are very minor, in order of 1.1- or 1.2-fold differences.

Robins: They are sometimes 1.5. It is a start. This kind of technology hasn't been fully applied yet.

Lovell-Badge: Until we know how much a subtle difference in gene expression can affect neuronal connectivity then we can't draw too many conclusions.

References

Aitken RJ, Marshall Graves JA 2002 The future of sex. Nature 415:963

Bishop CE, Whitworth DJ, Qin Y et al 2000 A transgenic insertion upstream of Sox9 is associated with dominant XX sex reversal in the mouse. Nat Genet 26:490–494

Carruth LL, Reisert I, Arnold AP 2002 Sex chromosome genes directly affect brain sexual differentiation. Nat Neurosci 5:933–934

Dewing P, Shi T, Horvath S, Vilain E 2003 Sexually dimorphic gene expression in mouse brain precedes gonadal differentiation. Brain Res Mol Brain Res 118:82–90

Hawley RS 2003 The human Y chromosome: rumors of its death have been greatly exaggerated. Cell 113:825–828

Koolhaas JM, Everts H, de Ruiter AJ, de Boer SF, Bohus B 1998 Coping with stress in rats and mice: differential peptidergic modulation of the amygdala-lateral septum complex. Prog Brain Res 119:437–448

Lim MM, Wang Z, Olazabal DE, Ren X, Terwilliger EF, Young LJ 2004 Enhanced partner preference in a promiscuous species by manipulating the expression of a single gene. Nature 429:754–757

Marshall Graves JA 2002 Evolution of the testis-determining gene — the rise and fall of SRY. In: The genetics and biology of sex determination. Wiley, Chichester (Novartis Found Symp 244) p 86–101

Maxson SC 1996 Searching for candidate genes with effects on an agonistic behaviour, offense, in mice. Behav Genet 26:471–476

Maxson SC, Roubertoux PL, Guillot P, Goldman D 2001 The genetics of aggression: from mice to humans. In: Martinez M (Ed) Prevention and control of aggression and the impact on its victims. Kluwer Academic, NY, p 71–81

Migeon CJ, Wisniewski AB, Gearhart JP et al 2002 Ambiguous genitalia with perineoscrotal hypospadias in 46,XY individuals: long-term medical, surgical, and psychosexual outcome. Pediatrics 110:e31

Thornhill AR, Burgoyne PS 1993 A paternally imprinted X chromosome retards the development of the early mouse embryo. Development 118:171–174

Wisniewski AB, Migeon CJ, Gearhart JP et al 2001 Congenital micropenis: long-term medical, surgical and psychosexual follow-up of individuals raised male or female. Horm Res 56:3–11

Xu J, Burgoyne PS, Arnold AP 2002 Sex differences in sex chromosome gene expression in mouse brain. Hum Mol Genet 11:1409–1419

Androgen receptor and molecular mechanisms of male-specific gene expression

Diane M. Robins

Department of Human Genetics, 4909 Buhl Building, University of Michigan Medical School, Ann Arbor, MI 48109-0618, USA

Abstract. Androgen's central role in male differentiation, fertility and aggression is evident from human pathologies and animal behaviour studies. Androgens directly regulate gene expression via the androgen receptor (AR), a member of the nuclear receptor superfamily. Nuclear receptors share a modular structure, with specialized domains for DNA binding, ligand binding, and transcriptional activation. Ligand-induced conformational changes in receptor trigger coregulators to modify chromatin structure. This in turn controls access of the transcriptional machinery to targeted genes. Given a common receptor structure and mode of action, AR must rely on several mechanisms for transcriptional specificity. As we have shown for the mouse sex-limited protein gene, these include use of divergent DNA binding sites, cooperativity between sites enhanced by intra- and intermolecular interaction of AR's N- and C-termini, and coactivator interactions. Although individual mechanisms lack sufficient specificity or strength, the summed interplay of response elements and accessory factor binding achieves precise gene activation. In addition to direct gene activation, AR elicits male-specific expression by modulating other hormonal pathways. For example, androgen control of growth hormone secretion induces male-specific genes in the liver. This sex-specific expression is further enforced by a novel class of KRAB zinc finger repressors. Their variation between species evidences diverging mechanisms of sexual differentiation, echoing AR's own recent origin.

2005 Molecular mechanisms influencing aggressive behaviours. Wiley, Chichester (Novartis Foundation Symposium 268) p 42–56

Steroid receptor action

That testosterone affects physical development, sexual differentiation and aggressive behaviour in males has been evident since Neolithic man began using castration in animal domestication more than 6000 years ago (Freeman et al 2001). Steroid hormones, including androgens (first purified from policemen's urine), were isolated early in the twentieth century as the biologically active compounds in several glandular extracts. By 1960, adrenal and gonadal steroids were proposed

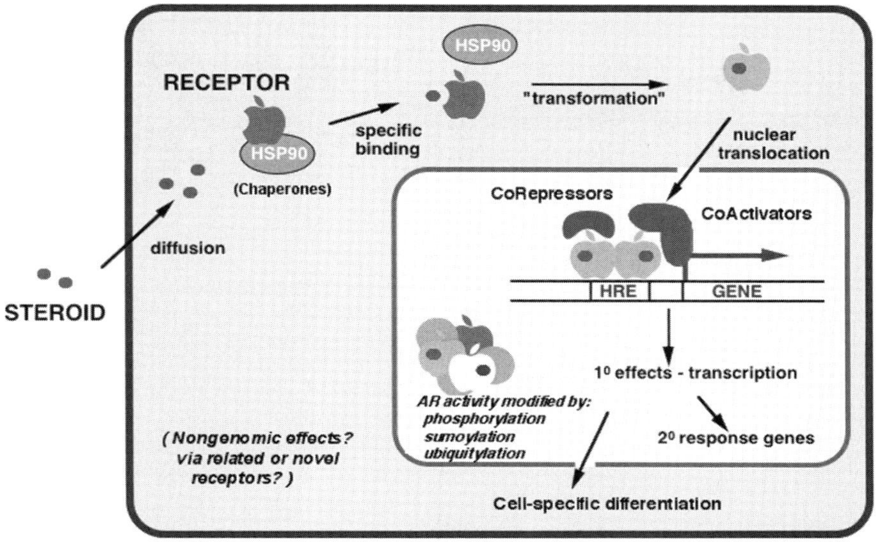

FIG. 1. Steroid hormone action. Steroids diffuse into cells and are retained by specific intracellular receptors. This binding causes conformational changes that release receptor from heat shock protein and chaperone complexes and allow import into the nucleus, where hormone response elements associated with target genes are bound. Transcription is the primary outcome of steroid receptor action and is affected by coactivator and corepressor proteins that modulate chromatin structure. See Beato (1989) for review.

to act at the level of gene regulation to influence cell proliferation, differentiation and homeostasis. Biochemical characterization of each steroid receptor led to a model of their mechanism, now visualized in molecular detail by the cloning of the receptor genes and analysis of their structure and function (Beato et al 1995).

In general, lipophilic steroid hormones diffuse into cells where they are bound with high affinity and specificity by intracellular receptors (Fig. 1). This binding induces a conformational change, releasing receptor from its association with chaperone molecules, notably heat shock proteins, and allowing translocation to the nucleus. In the nucleus, steroid receptors bind DNA sequences in or near their target genes, initiating a cascade of events leading to increased, or decreased, gene expression. Receptor structure reflects each separable function—there is a C-terminal hormone (ligand) binding domain (LBD), a central DNA binding domain (DBD), and a transcriptional activation function (AF1) within the large N-terminus (Fig. 2). Steroid receptors are in essence ligand-activated transcription factors (Beato et al 1995).

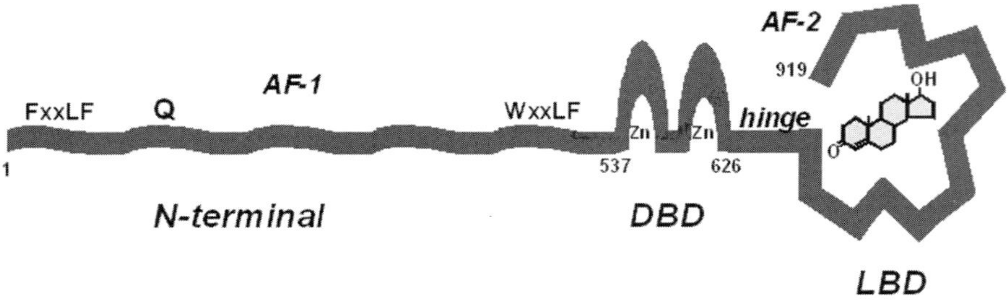

FIG. 2. Androgen receptor structure and functional domains. Structure of the large N-terminus has not yet been elucidated, but the crystal structure has been determined, in part by comparison to that of other nuclear receptors, for the DNA binding domain zinc fingers and the twelve α helices of the ligand binding domain (LBD) (see Buchanan et al 2001). Sites of some key functional elements are approximated on the diagram, including the FxxLF and WxxLF N-terminal regions that interact with the AF-2 domain in the LBD.

Glucocorticoid receptor (GR) was the first receptor cloned, followed rapidly by the oestrogen, progesterone, mineralocorticoid and androgen receptors (ER, PR, MR, AR). Comparison of the cDNAs revealed striking similarities, particularly in the two Cys_4 zinc-chelating fingers of the DBD. This domain served as a probe to identify not just other steroid receptors, but also a remarkably large family of previously unknown nuclear factors, numbering over 75 in the mammalian genome (Mangelsdorf et al 1995). The steroid receptors are a recently evolved, specialized class within the nuclear receptor superfamily. AR appears to be the most recent, perhaps reflecting rapid diversification of mechanisms for sexual differentiation. Whereas most superfamily members heterodimerize with the common partner, retinoid X receptor, the steroid receptors function primarily as homodimers. Steroid receptors are also distinct in their much larger N-terminal domains, suggesting differences in transcriptional activation mechanisms via AF1.

A key to understanding steroid receptor function, with broad application to transcription in general, came from studies addressing the critical role of the hormone in receptor activation — i.e. how do liganded and unliganded receptors differ? X-ray crystallography of several receptor ligand binding domains (LBDs) reveals a common structure consisting of 12 α helices and a β sheet (Moras & Gronemeyer 1998). A dramatic structural change occurs upon hormone binding, in which helix 12 flips up to appose helices 3, 4 and 5, encasing the ligand in what is called the mousetrap model. This creates a surface cleft that serves as a binding site for coactivator proteins that augment transcription. These coactivators account for the ligand-dependent transcription activation function (AF2) mapped to helix 12. Binding of antagonists rather than agonists shifts the position of helix 12, effectively blocking coactivator binding but allowing binding of corepressor proteins at another surface. The means by which these coactivators and corepressors transduce ligand binding to a transcriptional outcome resides in their ability to acetylate or deacetylate histones, either with their own intrinsic activity or via additional recruited proteins. By modifying nucleosome structure, coregulators promote (coactivator acetylation leading to chromatin decondensation) or prevent (corepressor deacetylation leading to compaction) access of the transcription machinery to the DNA. Thus, steroids regulate gene expression via their receptors by targeting chromatin modification to specific sites in the genome.

Mechanisms of androgen receptor transcriptional regulation

The steroid receptors are exquisitely specific with regard to their ligands, and activate distinct sets of target genes appropriate for disparate biological programs. Yet the *cis*-acting regulatory element first defined for GR also confers induction *in vitro* by PR, MR and AR (Beato 1989). We focused on this fundamental issue—how specific hormonal regulation of gene expression is attained when several receptors recognize similar DNA binding sites. We used the mouse sex-limited protein (*Slp*) gene as a model for male-specific expression, because of its accessibility to genetic, molecular and physiological manipulation (Robins 2004). Studies in this and other systems (Rennie et al 1993, Verrijdt et al 2003) highlight aspects of AR that diverge from the other receptors, particularly in DNA binding and in intramolecular and coregulator interactions (He & Wilson 2002, Heinlein & Chang 2002).

Slp was the first gene for which an androgen-specific response element was defined (Adler et al 1992). This enhancer, dubbed C′Δ9, resides 2 kb upstream of the start site of transcription and includes multiple receptor binding sites as well as binding sites for other transcription factors. Individual elements within the complex enhancer were characterized by functional transfection assays and protein–DNA interaction studies, focusing first on the role of the receptor binding sites (hormone response elements, or HREs) (Adler et al 1993, Ning & Robins 1999, Scheller et al 1996). The 3′ end of the enhancer is delimited by HRE-3, which adheres to the consensus sequence, an inverted repeat of 5′-TGTTCT-3′ half-sites separated by 3 bp. Immediately preceding HRE-3 is a weak consensus element (HRE-2) and an adjacent perfect half-site (HRE-1). HRE-3 is the only element that confers significant hormonal response in isolation, but response is not AR-specific and is obtained also for GR, MR or PR. HREs 1 and 2 show marginal activity on their own, but contribute within the context of the enhancer to both specificity and extent of response (Fig. 3). Specificity is also influenced in the context of the enhancer by mutations within HRE-3. Importantly, C′Δ9 has disparate activity in different cell types, indicating

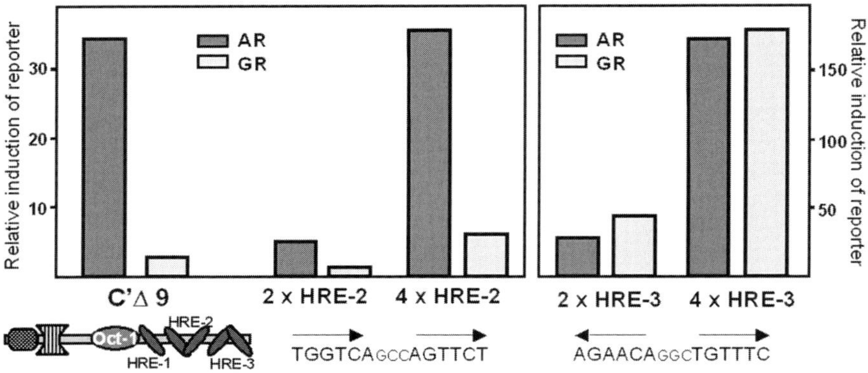

FIG. 3. Natural enhancers include specific and non-specific receptor binding sites and sites for non-receptor factors. The androgen-specific enhancer of the mouse *Slp* gene, C′Δ9, has three hormone response elements and binding sites for Oct-1, CBFa1 and factors still to be identified (Robins 2004). The histogram summarizes data from Adler et al (1992, 1993) and Verrijdt et al (2002), indicating relative hormonal induction of elements linked to reporter genes, activated by either androgen receptor (AR, dark bars) or glucocorticoid receptor (GR, light bars). The androgen-specific element HRE-2 is selective in response to androgen, but is very weak compared to the non-selective consensus HRE-3 (note different scales). In the context of the intact C′Δ9 fragment, single copies of each work in concert with additional factors to confer strong androgen-specific and cell-selective response.

that specificity is subject to the non-receptor factors present in the host. These early observations summed to the notion that while the HRE mediated hormonal activation, the nature of the response was dependent on the precise sequence of the response element and the array of neighbouring receptors and other transcription factor binding sites.

The complexity of receptor and cofactor interactions and their contextual dependence is underscored by recent work showing that AR, unlike other steroid receptors, can bind direct repeat sequences in an alternative head-to-tail dimer configuration (Reid et al 2001, Shaffer et al 2004). HRE-2 within the *Slp* enhancer has the structure of a direct repeat and binds AR specifically, but it relies on HRE-3 for functionality (Fig. 3; Verrijdt et al 2002). At least in some androgen-specific enhancers thus far characterized, these two classes of consensus and non-consensus elements cooperate to mediate androgen specificity. Interactions may be sequential, with AR first targeted to highly specific sequences, allowing cooperative recruitment of additional ARs to consensus receptor binding sites, and augmentation of activity by cell-specific interactions with neighbouring transcription factors.

Multiple lines of evidence demonstrate the importance of AR cooperativity, perhaps the most significant being the crucial role of weak HREs in hormonal response. Even though HRE-2 has little independent activity, its mutation within the context of the *Slp* enhancer has as deleterious an effect as mutation of the consensus HRE-3 (Scheller et al 1996). Additionally, HRE-1 can bind receptor only if HRE-3 is also present, and influences hormone-specific response since GR does not bind this site. The weak binding sites may enhance sensitivity to varying receptor (and hormone) concentrations, allowing response to reflect varying physiological conditions.

The molecular basis for this cooperativity relies at least in part on AR's unusual dependence on direct intra- and intermolecular contact between its N- and C-termini, known as N/C interaction (He et al 1999). N/C interaction is induced by ligand binding, and is critical for receptor stabilization and transcriptional function. This interaction likely stabilizes AR monomers in an antiparallel orientation within the dimer. Elegant work from Elizabeth Wilson's laboratory has shown that conserved motifs in the very N-terminus (^{23}FXXLF^{27}) and just before the DBD (^{433}WXXLF^{437}) (see Fig. 2) bind to AF-2 in N/C interaction, with the FXXLF motif having the prominent role (He et al 2000). Interestingly, the FXXLF motif competes with the LXXLL sequence of the p160 coactivators for binding to AF-2. This may underlie AR's evolution of alternative coactivator interactions that rely on N-terminal rather than LBD sequences (Bevans et al 1999).

In addition to co-regulators that interact with other nuclear receptors, exemplified by the p160 family member SRC-1, proteins that selectively interact with AR have been sought (Heinlein & Chang 2002). Several factors have been isolated that preferentially accentuate AR activity, but few of these interactions are exclusive. Interestingly, several accessory factors modulate AR activity via other mechanisms (AR stability, modification, nuclear import or export). In addition, some transcription factors frequently cooperate with AR, such as HNF-3α (FOXA1) that co-regulates androgen-dependent genes in the prostate (Gao et al 2003).

Intriguing evidence for an as-yet unidentified factor that interacts with N-terminal sequences of AR comes from studying AR's own androgen-specific enhancer (Grad et al 2001). This enhancer contains four consensus HREs distributed over two exons. Sequences within a 350 bp cDNA fragment drive androgen-specific induction of a reporter gene. Chimeric AR/GR receptors (Scheller 1998) reveal that specificity is derived from only the N-terminal domain. The DBD of either receptor provides target recognition, and AR N/C interaction is also not essential.

While AR's intragenic enhancer appears to rely on a factor that is present in only some cell types, more general factors can be critical components of specific response. For *Slp*, a key co-regulator appears to be the ubiquitous transcription factor Oct-1. Its enhancer binding site is occupied *in vivo* in a hormone-dependent and tissue-specific manner (Scarlett & Robins 1995). Oct-1 is known to interact with GR, positively or negatively, depending on the arrangement of their binding sites (Wieland et al 1991). We found in protein interaction studies that binding of Oct-1 and receptors to their cognate DNA sites promoted interaction of AR, but not GR, with Oct-1 (Gonzalez & Robins 2001). DNA binding caused a detectable conformational change in AR's LBD, which depended on N/C interaction, and increased recruitment of SRC-1 to the AR-Oct-1 complex. Differential DNA-dependent protein–protein interactions may thus direct a specific transcriptional response, even though the factors and co-regulators eliciting such a response are not in themselves inherently discriminatory.

Enhancers such as that of *Slp* provide insight into the complexity of AR action and how it differs from that of other nuclear receptor family members. Particular modes of DNA recognition, N/C interaction, interaction with other transcription factors and co-regulators conspire towards precise gene activation that is highly dependent on cell and promoter context. Each point of interaction is a potential target for modulation of AR activity. While several syndromes, such as androgen insensitivity, indicate abnormal AR function, the single greatest health problem associated with AR is prostate cancer. Prostate tumours are androgen dependent in their initiation, and AR-dependent (but hormone-independent) in their progression. The quest for treatment of this disease has promoted intense scrutiny of AR action. Studies of prostate cancer have also necessitated examining broader aspects of androgen action, in which testosterone is a precursor as well as a hormone and regulates gene expression indirectly as well as directly. Further, male-specific gene expression, in prostate and in other tissues, can be enforced by mechanisms that do not depend on androgens.

Indirect AR effects

Unique physiological understanding of AR structure and function is possible because the *Ar* gene is located on the X chromosome. The effect of any mutation is therefore evident in males, with defects ranging from subtly reduced AR activity causing partial androgen insensitivity (PAIS) to infertile XY females resulting from completely null alleles (CAIS, or testicular feminization, *tfm*) (Quigley et al 1995, Buchanan et al 2001). These patients, plus an array of natural and engineered mouse mutants, reveal indirect as well as direct effects of AR on gene expression. Some of these effects are due to interaction at multiple levels with oestrogen receptor (ER) regulation. Testosterone plays a critical role in hormonal balance, in both sexes, being a precursor not only for the more potent androgen dihydrotestosterone, but also for oestradiol, by virtue of aromatase action. In cell-specific manners, androgens and oestrogens balance each other's actions, by impacting ligand synthesis, receptor expression, or myriad steps in opposing pathways.

To assess the relationship of sex steroids in numerous tissues, we created mice deficient in both AR and ER by crossing females carrying the classic *tfm* allele with ERα knockout males. F2 male, female, or *tfm*/Y progeny, with or without ERα (six genotypes in all), were evaluated, most extensively for skeletal parameters (Tozum et al 2004). Loss of ERα decreased femur length in males, but increased it in females, suggesting a reversal of sex steroid-dependent skeletal dimorphism in the absence of ERα. Interestingly, *tfm* mice had the longest bones, and lengths were reduced by lack of ERα, similar to the male. In contrast, bone mineral density (BMD) decreased in males in the absence of ERα, whereas female and *tfm* BMD were unaffected. This suggests that the enhancing or opposing actions of the sex steroids are cell and tissue specific, and complicated by the fact that *tfm* mice are male-like in some parameters and female-like in others. Clarification may come from the 'floxed' AR allele, allowing its deletion in specific tissues at specific times, as well as the creation of an AR−/AR− female to contrast to *tfm*/Y (Yeh et al 2003). This should distinguish systemic from cell-autonomous AR action effects.

Sex-specific differences in gene expression are common in non-reproductive as well as reproductive organs, and a recent study from our laboratory delineates one mechanism by which this occurs in liver (Tullis et al 2003, Krebs et al 2003). A wide variety of liver proteins are expressed in sexually dimorphic patterns, including many steroid and drug metabolizing cytochrome P450s and several serum proteins, particularly pheromone carriers and complement factors. Hormonal induction of these proteins differs in males and females due to sex-specific patterns of growth hormone (GH) release, caused by gonadal steroid action on the hypothalamo-pituitary axis (Fig. 4). Pulsatile GH signalling in males, in contrast to the more continuous levels produced in females, is transduced in liver by STAT5b, which upon activation enters the nucleus and drives male-specific gene transcription (Waxman 2000). An independent axis that greatly accentuates hepatic sex differences is revealed in mice carrying variant *Regulator of sex-limitation* (*Rsl*) alleles. Mutant *rsl* females aberrantly express high levels of male-specific proteins, which are also higher in *rsl* compared to wild-type males. The *rsl* regulation occurs in *tfm* or hypophysectomized mice, indicating its lack of dependence on either androgen or GH (Tullis et al 2003).

We positionally cloned *Rsl*, which proves to be a pair of Kruppel-associated box zinc finger proteins (KRAB-ZFPs) that account for the *Rsl* phenotype by repressing male-specific liver genes in both sexes (Krebs et al 2003). We found that *Rsl* maps to a cluster of 24 of these KRAB-ZFP genes on mouse chromosome 13. The recently duplicated *Rsl1* and *Rsl2* harbour mutations in *rsl* mice, and, encompassed in separate bacterial artificial chromosomes, each rescues

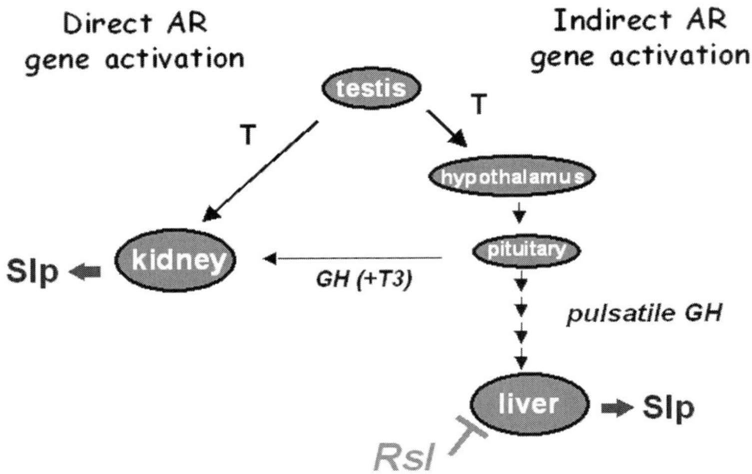

FIG. 4. Male-specific gene expression can be conferred by different mechanisms. In the kidney, the mouse *Slp* gene is directly activated by androgen and AR, with some GH and thyroid hormone input. The same gene is activated in liver at puberty by testosterone acting through the hypothalamo-pituitary axis to cause pulsatile GH release. In addition, the *Rsl* KRAB zinc finger proteins repress the male-specific genes in the liver in both sexes, independently of androgen or GH, to accentuate sex-specific gene expression (Tullis et al 2003, Krebs et al 2003).

correct regulation of a distinct subset of the male-specific target genes. That is, *Rsl1* restores sex-specific expression of *Slp* and *Cyp2d9*, while *Rsl2* represses mouse urinary proteins (MUPs). These paralogous genes apparently have diverged functionally and now act in concert to enforce liver dimorphism.

Rsl1 and *Rsl2* are the first KRAB-ZFPs for which biological roles have been assigned. This is remarkable since there are hundreds of KRAB-ZFPs in both human and mouse genomes, with known ability to repress transcription (Mark et al 1999). *Rsl* orthologues exist in rat but are not obvious in man, suggesting that KRAB-ZFPS generally may participate in species-specific functions. Species-specific genes in mammals are most often differentially amplified members of multigene families that function in reproduction, immunity, or xenobiotic metabolism. Since Rsl's targets fall into these categories, it is tempting to speculate that regulators of these genes may also be species-specific. Perhaps by diversifying gene expression patterns, KRAB-ZFPs have acquired a significant role in speciation and evolution.

In summary, androgen regulates male-specific gene expression directly via AR, which is a hormone-activated transcription factor, and indirectly by controlling effect (as for ER) or secretion (as for GH) of other hormonal pathways. Furthermore, sex-specific differences in gene expression can be elicited independent of sex steroids, as demonstrated by discovery of the *Rsl* repressors. That numerous mechanisms exist for sex-specific gene expression, and that a surprisingly large number of genes are expressed in sex-specific patterns, underscores the importance of sex-specific gene expression. Many of these differences in gene expression patterns impact reproduction, with diversification in expression patterns leading to speciation and thus providing a driving force in evolution.

Acknowledgement

This work was funded by the NIH. I thank all the members of the lab, past and present, for their contributions, their insights, and their camaraderie.

References

Adler AJ, Danielsen M, Robins DM 1992 Androgen-specific gene activation via a consensus glucocorticoid response element is determined by interaction with nonreceptor factors. Proc Natl Acad Sci USA 89:11660–11663

Adler AJ, Scheller A, Robins DM 1993 The stringency and magnitude of androgen-specific gene activation are combinatorial functions of receptor and nonreceptor binding site sequences. Mol Cell Biol 13:6326–6335

Beato M 1989 Gene regulation by steroid hormones. Cell 56:335–344

Beato M, Herrlich P, Schutz G 1995 Steroid hormone receptors: many actors in search of a plot. Cell 83:851–857

Bevans CL, Hoare S, Claessens F, Heery DM, Parker MG 1999 The AF1 and AF2 domains of the androgen receptor interact with distinct regions of SRC1. Mol Cell Biol 19:8383–8392

Buchanan G, Greenberg NM, Scher HI, Harris JM, Marshall VR, Tilley WD 2001 Collocation of androgen receptor gene mutations in prostate cancer. Clin Cancer Res 7:1273–1281

Freeman ER, Bloom DA, McGuire EJ 2001 A brief history of testosterone. J Urol 165:371–373

Gao N, Zhang J, Rao MA et al 2003 The role of hepatocyte nuclear factor-3α (Forkhead Box A1) and androgen receptor in transcriptional regulation of prostatic genes. Mol Endocrinol 17:1484–1507

González MI, Robins DM 2001 Oct-1 preferentially interacts with androgen receptor in a DNA-dependent manner that facilitates recruitment of SRC-1. J Biol Chem 276:6420–6428

Grad JM, Lyons LS, Robins DM, Burnstein KL 2001 The androgen receptor (AR) amino-terminus imposes androgen-specific regulation of AR gene expression via an exonic enhancer. Endocrinol 142:1107–1116

He B, Wilson EM 2002 The NH_2-terminal and carboxyl-terminal interaction in the human androgen receptor. Mol Gene Metab 75:293–298

He B, Kemppainen JA, Voegel JJ, Gronemeyer H, Wilson EM 1999 Activation function 2 in the human androgen receptor ligand binding domain mediates interdomain communication with the NH_2-terminal domain. J Biol Chem 274:37219–37225

He B, Kemppainen JA, Wilson EM 2000 FxxLF and WxxLF sequences mediate the NH_2-terminal interaction with the ligand binding domain of the androgen receptor. J Biol Chem 275:22986–22994

Heinlein C, Chang C 2002 Androgen receptor (AR) coregulators: an overview. Endocr Rev 23:175–200

Krebs C, Larkins L, Price R, Tullis K, Miller RD, Robins DM 2003 *Regulator of sex-limitation* (*Rsl*) encodes a pair of KRAB zinc finger genes that control sexually dimorphic liver gene expression. Genes & Dev 17:2664–2674

Mangelsdorf DJ, Thummel C, Beato M et al 1995 The nuclear receptor superfamily: the second decade. Cell 83:835–839

Mark C, Abrink M, Hellman L 1999 comparative analysis of KRAB zinc finger proteins in rodents and man: Evidence for several evolutionarily distinct subfamilies of KRAB zinc finger genes. DNA Cell Biol 18:381–396

Moras D, Gronemeyer H 1998 The nuclear receptor ligand-binding domain: structure and function. Curr Opin Cell Biol 10:384–391

Ning YM, Robins DM 1999 AML3/CBFα1 is required for androgen-specific activation of the enhancer of the mouse sex-limited protein (*Slp*) gene. J Biol Chem 274:30624–30630

Quigley CA, De Bellis A, Marschke KB, El-Awady MK, Wilson EM, French FS 1995 Androgen receptor defects: Historical, clinical and molecular perspectives. Endocr Rev 16:271–321

Reid K, Hendy S, Saito J, Sorensen P, Nelson C 2001 Two classes of androgen receptor elements mediate cooperativity through allosteric interactions. J Biol Chem 276:2943–2952

Rennie PS, Bruchovsky N, Leco KJ et al 1993 Characterization of two cis-acting DNA elements involved in the androgen regulation of the probasin gene. Mol Endocrinol 7:23–36

Robins DM 2004 Multiple mechanisms of male-specific gene expression: lessons from the mouse sex-limited protein (*Slp*) gene. Prog Nucl Acids Res 78:1–36

Scarlett CO, Robins DM 1995 *In vivo* footprinting of an androgen-dependent enhancer reveals an accessory element integral to hormonal response. Mol Endocrinol 9:413–423

Scheller A, Scheinman RI, Thompson E, Scarlett CO, Robins DM 1996 Contextual dependence of steroid receptor function on an androgen-responsive enhancer. Mol Cell Endocrinol 121:75–86

Scheller A, Hughes E, Golden KL, Robins DM 1998 Multiple receptor domains interact to permit, or restrict, androgen-specific gene activation. J Biol Chem 273:24216–24222

Shaffer PL, Jivan A, Dollins DE, Claessens F, Gewirth DT 2004 Structural basis of androgen receptor binding to selective response elements. Proc Natl Acad Sci USA 101:4758–4763

Tozum TF, Oppenlander ME, Koh-Paige AJ, Robins DM, McCauley LK 2004 Effects of sex steroid receptor specificity in the regulation of skeletal metabolism. Calcif Tissue Int 75: 60–70

Tullis K, Krebs C, Leung J, Robins DM 2003 The regulator of sex-limitation gene, *Rsl*, enforces male-specific liver gene expression by negative regulation. Endocrinol 144:1854–1860

Verrijdt, Haelens A, Claessens F 2003 Selective DNA recognition by the androgen receptor as a mechanism for hormone-specific regulation of gene expression. Mol Gene Metabol 78: 175–185

Verrijdt G, Schwauwaers K, Haelens A, Rombauts W, Claessens F 2002 Functional interplay between two response elements with distinct binding characteristics dictates androgen specificity of the mouse sex-limited protein enhancer. J Biol Chem 277: 35191–35201

Waxman DJ 2000 Growth hormone pulse-activated STAT5 signaling: a unique regulatory mechanism governing sexual dimorphism of liver gene expression. In: Mechanisms and biological significance of pulsatile hormone secretion. Wiley, Chichester (Novartis Found Symp 227) p 61–74

Wieland S, Dobbeling U, Rusconi S 1991 Interference and synergism of glucocorticoid receptor and octamer factors. EMBO J 10:2513–2521

Yeh S, Hu Y-C, Wang P-H et al 2003 Abnormal mammary gland development and growth retardation in female mice and MCF7 breast cancer cells lacking androgen receptor. J Exp Med 198:1899–1908

DISCUSSION

Pfaff: In the field demonstrating brain mechanisms controlling sex behaviours, people are beginning to be able to dissect the individual coactivators for steroid-influenced sex behaviours. As you think about androgen actions and aggressive behaviour, if someone walked you to the end of a plank and you had to design a behavioural experiment, which coactivators would be good candidates?

Robins: I think all of them. Certainly, SRC-1, -2 and -3 are very important for the steroid receptor interactions. One of the problems that has come up with knockouts is that in the absence of one there may be compensation by the others, so we always have to try to control for this. People have looked for proteins such as coactivators that are specific to AR. I don't think any of them are enormously specific, but rather they are contextually dependent: they will interact with other proteins. AR finds more than one way to do any particular thing. The knockouts will be informative, but I am sure that there is more than one way to get any particular outcome.

Pfaff: We talked about the various molecular relationships, of which there were several types, each of which has a certain amount of variability associated with it. I was thinking about the individual variability of behaviour within a species, even androgen-stimulated behaviour. Within the results reported in your talk, we see the basis for animal–animal variation in terms of response to androgen hormone. In fact, it is surprising that behaviour isn't more variable than it is.

Robins: As someone who doesn't do behaviour studies, I think it is quite variable.

Suomi: As one way to think of contextually dependent behaviour, would differences in developmental context also be possible?

Robins: I think that is pretty clear. There are classic studies where you can alter the behaviour of animals after puberty by prenatal exposure to testosterone or oestrogen. Presumably, at a molecular level one of the things this is doing is imprinting gene expression to be regulated differentially later. You clearly can affect embryos this way. I have always enjoyed this 'nearest neighbour effect' in mouse development, where behaviour of a mouse varies depending on whether they were next to a female pup or male pup *in utero*.

Keverne: Are any of those *Rsl* genes which act as transcriptional repressors monoallelically expressed? Are they found in imprinting clusters?

Robins: This has not been looked at, but they may be having these subtle effects. We would never have found *Rsl* if we didn't have a mouse mutant, and the mutant mice are perfectly fine. There are actually two behavioural correlates of variation in *Rsl* genes. These repressors regulate the MUP pheromone carriers, so they are overexpressed in the mutants. I thought that this would make the females unattractive, because they are overexpressing male proteins in their urine and they are probably marking territory without realizing it. We did a mate choice test using scented bedding. It turns out that our females are 50% more attractive to the males than the wild-type females. Chris Krebs in my lab suggested that there may be more of the pheromone carrier, but the females are still expressing female pheromones, kind of like wearing too much perfume. The effect for the females of being exposed to males that are overexpressing MUPs is that they may go through early puberty. Another interesting aspect of this story from an evolutionary point of view is that the mutation in *Rsl1* goes back to *Mus domesticus* and is present in all inbred strains, but *Rsl2* shows different mutations in different strains. It is a more recent mutation, and there seems to be selection such that once *Rsl1* is mutated *Rsl2* is also lost. The selection may be conferred by the caretakers at the Jackson lab, since they look for mice that breed early and often. It is advantageous to go through early puberty if you are a mouse in the Jackson lab, but perhaps not in the wild.

Keverne: When you say you are getting overexpression of MUPs, is this overexpression of female MUPs or is this expression of the whole MUP family?

Robins: We are only probing for three male-specific MUPs. The MUP gene family has about 35 members. Some are liver specific, some are lachrymal gland specific, and so on. We are looking at three of the liver MUPs that are enhanced in expression in male. The females make the same ones but perhaps at one-eighth of the male level. The carriers have less directional capacity since they will bind whatever pheromones are present to retard their volatilization in urine.

Keverne: That is one possibility. Another is that the MUPs might themselves be binding to the V2R receptor.

Robins: Yes, and to some extent all degraded proteins make it through into urine in mice. There is a degree of mate choice based on H2 haplotypes, for example. The different protein profiles may have an effect in the wild.

Dulac: According to your selection theory, why wouldn't females have higher MUP levels? This would make them more attractive.

Robins: The females do have higher MUP levels — both sexes make more of the male-specific proteins. But the females probably still synthesize female pheromones regardless of the carrier. With early puberty, my guess is that it is the exposure to male pheromone, not the MUPs *per se*, which is important.

Dulac: But in the wild, why wouldn't you expect the female to increase MUP levels?

Robins: I don't know that this has been found in the wild. *Rsl2* in the wild looks wild-type in the one *Mus domesticus* mouse we have looked at.

Pfaff: I was going to ask Craig about androgenic effects on vasopressin expression in the brain. First, though, I have a couple of questions that seem to contradict each other, but they don't really. The first has to do with the fact that androgen effects on aggressive behaviour usually seem to take a long time. Have there been microarray experiments done under such conditions that one can see not only a primary wave of androgen-stimulated gene expression, but possibly also a secondary or tertiary wave? This would be somewhat similar to what has been done with the yeast cell cycle. Such microarray data would provide a feast for us neurobiologists, who then would start dreaming about which hormone-influenced genes could be important.

Suomi: What do you mean by 'a long time'?

Pfaff: In rats, if I were looking at an effect of testosterone on aggressive behaviour, I wouldn't want to do the assay any quicker than two weeks.

Robins: Much of the work being done now is on prostate cancer, since that is where funding is available. A lot of microarray work has been done looking at timing of gene expression. From these prostate studies it seems that there may be ligand-independent activation of AR. It is subject to all kinds of phosphorylation, ubiquitination and sumoylation, just like oestrogen receptor. I don't know whether this analysis has been applied to behaviour.

Pfaff: I don't think it has. This leads into my second question. With oestrogens and thyroid hormones one of the huge current stories has to do with these 'rapid actions' in which so-called nuclear receptors seem to be having a second career out at the membrane. Is this true of androgens as well?

Robins: Yes, AR is following right along.

Pfaff: Would it be the case, then, that AR operating near the cell membrane might transduce some kind of effect which ultimately results in a change in transcription, thus affecting behaviour?

Robins: This is being looked at in the context of the switch from androgen-dependent to androgen-independent disease in prostate cancer. Maybe one can now take prostate cancer as a molecular model for aggression studies.

Pfaff: I think you said that this long AR N-terminus makes additional opportunities for these ligand-independent interactions.

Robins: Yes.

Pfaff: Craig Ferris, what does this mean for vasopressin neurobiology in the brain?

Ferris: I'll put it in its historical context. In the late 1970s David de Wied from the Netherlands was the first to show that vasopressin the neurohormone released from the pituitary gland could act on the brain to alter cognitive behaviour. From de Wied's early work came a generation of Dutch scientists that thoroughly investigated the neurobiology of vasopressin as a chemical signal in the brain altering behaviour across multiple species. Geert deVries made the serendipitous discovery that castration eliminated vasopressin immunostaining in limbic brain areas suggesting gene expression was androgen sensitive. Subsequent work in their own and other labs showed it was mostly responsive to estrogens and not testosterone. Indeed, vasopressin neurons in limbic areas colocalize oestrogen receptors. As more and more labs got involved in vasopressin it became very apparent that there was a great deal of variability between species. For example, there is an abundance of vasopressin fibres in the septum of rat and mouse, but very little in the hamster. The distribution of fibres and the extrahypothalamic sources of neuropeptide are not consistent across species. If you look at rhesus and humans they are more like the hamster than the mouse or rat.

Pfaff: Those of us who are interested in comparing hormone-dependent behaviours across species will sooner or later have to know what the difference is between a paralogue, homologue and an orthologue.

Robins: ERα in the mouse is an orthologue of ERα in humans. It is a one-to-one cross species gene comparison. A homologue would be ERβ compared with ERα, because it is similar, but different in precise function. Paralogues arise from a gene duplication event. Usually people refer to a gene as a paralogue if it has already moved to another chromosome. The homeobox clusters where there are homologous genes in a row exist in four paralogous clusters in mammals, A, B, C, D. The genes in the same positions within these clusters would be paralogues. There are many similarities in their regulation and function, but they have distinctions and are now evolving separately. In the *Rsl* cluster it looks as if we had an inverted duplication of four genes: the two that are in the same position and are most similar to each other by sequence are the most recently duplicated pair, and I would refer to them as paralogues.

Quantitative trait locus analysis of aggressive behaviours in mice

Edward S. Brodkin

Center for Neurobiology and Behavior, Department of Psychiatry, University of Pennsylvania School of Medicine, 415 Curie Boulevard, Room 111, Philadelphia, PA 19104-6140, USA

Abstract. Quantitative trait locus (QTL) analysis is a forward genetic strategy (phenotype to gene) that enables an investigator to start with a phenotype of interest (e.g. aggressive behaviours), and then identify chromosomal regions (QTLs) and, ultimately, specific gene variants (alleles) within those regions, that influence quantitative variation in the phenotype. QTL analysis is an important complement to reverse genetic strategies (gene to phenotype), such as the generation of knockout and transgenic mice. Although a propensity for aggressive behaviours is partially heritable in mice and other mammals, very few aggression QTLs have been mapped. This slow progress is likely due, in part, to the complexity of aggressive behaviours as phenotypes, which are affected by many non-genetic (environmental and random) factors and gene-environment interactions. This paper reviews the general principles of QTL analysis, as well as the non-genetic factors that can confound aggression QTL studies. Some examples of successfully mapped intermale mouse aggression QTLs are presented, such as QTLs on chromosomes 10 and X. Strategies for fine mapping these loci are discussed, and candidate genes are considered. Finally, newly available mouse genetic resources that may facilitate QTL analysis of aggressive behaviours are suggested, such as consomic mouse strains.

2005 Molecular mechanisms influencing aggressive behaviours. Wiley, Chichester (Novartis Foundation Symposium 268) p 57–77

This paper will address quantitative trait locus (QTL) analysis of mouse aggressive behaviours, including the following topics:

- the general principles of QTL analysis
- non-genetic factors that affect aggression QTL studies
- progress to date toward the identification of mouse aggression QTLs, and
- newer mouse genetics resources for mapping aggression QTLs.

This paper will not cover studies of aggression in induced mutants (transgenes, knockouts), as the latter topic has been reviewed recently elsewhere (Maxson & Canastar 2003, Miczek et al 2001). This paper will focus on intermale offensive aggression, which has been the subject of most published aggression QTL studies.

Principles of quantitative trait locus (QTL) analysis

Definitions

In contrast to Mendelian traits, each of which is controlled primarily by a single gene and shows a simple pattern of inheritance (e.g. autosomal dominant, autosomal recessive, or X-linked), quantitative traits are each affected by multiple genes and environmental factors and show complex patterns of inheritance. Quantitative traits are usually measured as continuous variables on a numerical scale (e.g. number of seconds spent in a particular behaviour). Chromosomal regions containing a gene (or genes) that affect the level of a quantitative trait are called quantitative trait loci (QTLs). The relevant genes in these regions have been called quantitative trait genes (QTGs) (Hitzemann et al 2003). Quantitative trait locus (QTL) analysis is an experimental strategy for identifying QTLs, and ultimately QTGs, that affect quantitative traits. Because of the complexity of these traits, progress in identifying QTGs has been slow compared to that in cloning genes underlying Mendelian traits (Glazier et al 2002).

Coarse mapping of QTLs

The initial goal of a typical QTL mapping study is to identify chromosomal regions that contribute to differences between two inbred mouse strains in a phenotype of interest. Inbred strains are produced by 20 consecutive generations of brother–sister mating. All same-sex members of an inbred strain are genetically identical (with the exception of occasional spontaneous mutations that occur in the genome) and are homozygous at every locus throughout the genome, i.e. have two identical copies of every gene in the genome. The first steps in QTL analysis are to identify two inbred strains that differ in a phenotype of interest, and then to crossbreed the strains to produce reciprocal F1 (first filial) generation hybrids. In one set of hybrids, a female from the strain 1 is mother, and a male from the inbred strain 2 is father; and in the reciprocal set of hybrids, the converse breeding scheme is used (mother from inbred strain 2; father from inbred strain 1). The phenotype of the F1 hybrids is compared to those of the parental inbred strains to reveal dominance or semi-dominance relationships between the alleles that affect the phenotype. Phenotypic differences between reciprocal F1 hybrids indicate that one or more of the following factors may affect the trait: (1) sex linkage (X- or Y-linked traits), (2) genomic imprinting of QTLs that affect the phenotype, (3) prenatal maternal effects (effects of intrauterine environment), and/or (4) postnatal maternal or paternal effects (effects of maternal and/or paternal parenting behaviour on offspring).

Next, the F1 hybrids are either intercrossed or backcrossed to one of the parental inbred strains, to produce an intercross (F2) or backcross (N2) population (see Figs

1 and 2). Relative advantages and disadvantages of an intercross vs. a backcross breeding strategy are reviewed elsewhere (Darvasi 1998, Silver 1995). A population of at least several hundred mice is usually necessary to achieve sufficient statistical power to detect QTLs (Darvasi 1998). The phenotype is measured in all members of the F2 or N2 population, each of which possesses a unique combination of genetic material from the two progenitor inbred strains. By genotyping markers spaced throughout the genome, one determines whether the two copies of genetic material at each marker came from one parental inbred strain (homozygous from parental strain 1), from the other strain (homozygous from parental strain 2), or from both strains (heterozygous) (see Figs 1 and 2). A full genome scan involves genotyping markers spaced at approximately 20–30 centiMorgan (cM) intervals, which should provide full coverage of the genome, except for the non-pairing (male-specific) region of the Y chromosome (Silver 1995).

Statistical tests are carried out to determine whether the genotype at each marker affects quantitative levels of the trait in the intercross or backcross populations. Because at least 100 markers are genotyped in a full genome scan, a correction for multiple testing must be made to the significance threshold (Doerge & Churchill 1996, Lander & Kruglyak 1995). Several software packages are available for carrying out this statistical analysis, including MAPMAKER QTL (*http://www.broad.mit.edu/genome_software/*) and QTX (*http://www.mapmanager.org/*) (Manly & Olson 1999).

From coarse to fine mapping and candidate gene analysis

By identifying genetic markers that show significant linkage to the trait, as described above, one maps each QTL to a large chromosomal region ($\sim$10–30 cM) ('coarse mapping'), that, in most instances, contains hundreds of genes. To proceed towards identification of the relevant gene(s) in the region, one carries out breeding crosses that generate more chromosomal recombinations in the region, using any of several methods, including the generation of interval-specific congenic strains (Darvasi 1997). Measuring the phenotype in these strains can narrow the region of interest and greatly shorten the list of positional candidate genes. One can further shorten this list by picking candidate genes that are expressed in tissues likely to influence the traits of interest (Su et al 2004). These candidate genes are then sequenced in the two parental inbred strains looking for sequence differences in coding or regulatory regions.

After fine mapping the QTL interval and shortening the list of plausible candidate polymorphisms, the major challenge remains—proving definitively which nucleotide polymorphism underlies the QTL. The most direct proof would be replacing one strain's allele with another strain's allele (creating a

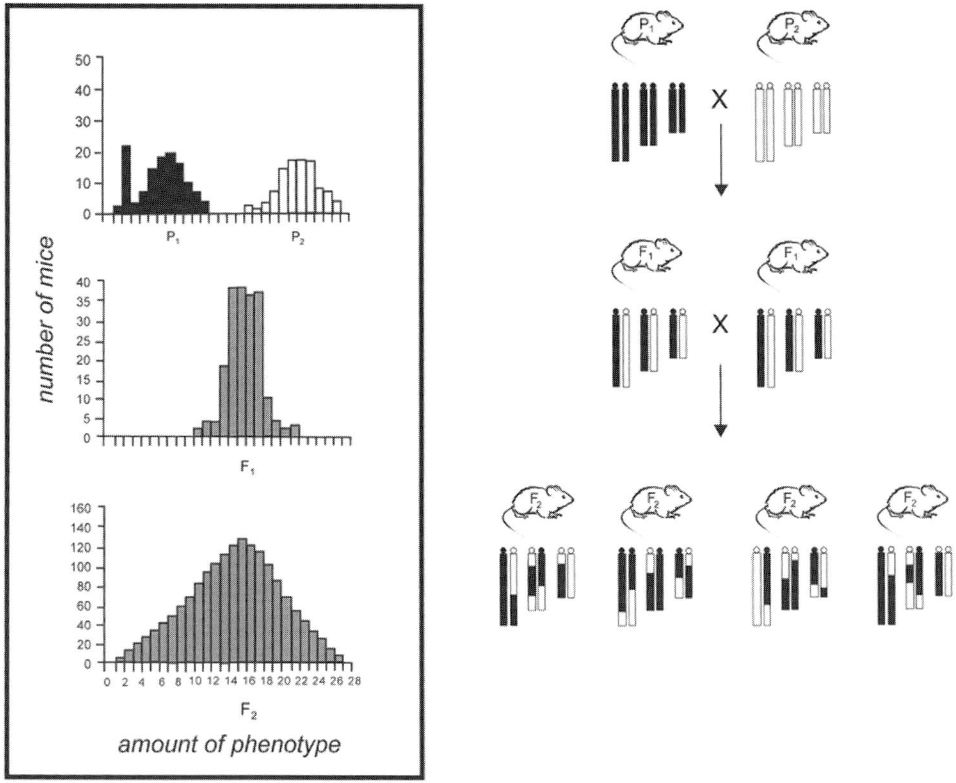

FIG. 1. Intercross breeding strategy for mapping quantitative trait loci (QTLs). On the right, the parental, F1 hybrid, and intercross (F2) mouse generations are depicted. The two parental strains (P1 and P2) are crossed to produce F1 hybrids, which are then intercrossed to produce F2 mice. For each animal, three pairs of homologous chromosomes are depicted (mice have a total of 20 pairs of homologous chromosomes). Chromosomes from parental strain 1 (P1) are represented in black, and chromosomes from parental strain 2 (P2) are represented in white. Each F2 mouse has a unique combination of P1 and P2 genes. The box on the left shows idealized phenotype histograms for the three generations.

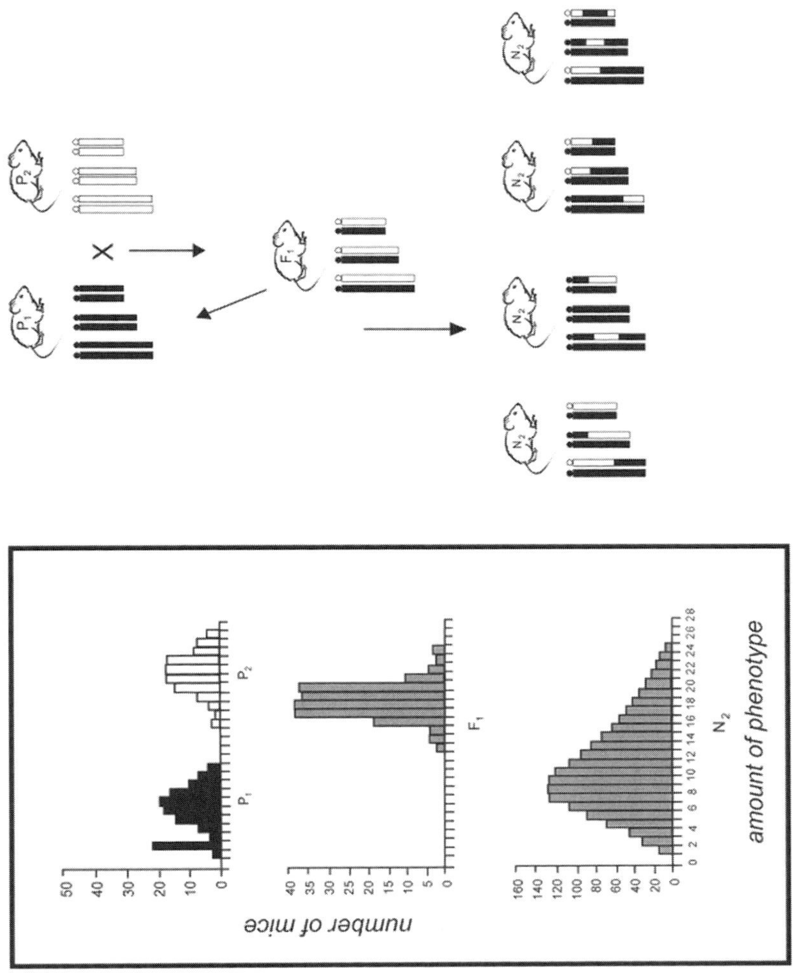

FIG. 2. Backcross breeding strategy for mapping QTLs. In this breeding strategy, the F1 hybrid is backcrossed to one of the parental strains, in this case, to parental strain 1 (P1).

knock-in mouse) and thus changing the phenotype in the expected direction. Other lines of evidence may provide a convincing preponderance of data implicating a candidate gene (see Abiola et al 2003).

Usefulness of QTL analysis in the context of other genetic approaches

A single QTL mapping experiment using two inbred progenitor strains will not identify all genes that affect the phenotype—it can only identify the subset of alleles that accounts for the phenotype difference between the two particular progenitor strains. Additional genes could potentially be identified by carrying out QTL analyses using other inbred strains. More broadly, QTL analysis is one among a wide array of genetic methodologies for dissecting complex traits. These methodologies include both 'forward genetic' strategies for identifying genes underlying phenotypes of interest (e.g. QTL analysis; ENU mutagenesis; study of spontaneous mutants), as well as 'reverse genetic' strategies for studying the effect of a known gene mutation on various phenotypes (e.g. analysis of phenotypes in knockout and transgenic mice). These genetic strategies are complementary, in fact, interdependent (Belknap et al 2001). For example, the effects of a gene mutation (e.g. a gene knockout) on a phenotype (revealed by a reverse genetics experiment) will often vary depending upon the inbred strain genetic background, that is, due to the effect of modifier loci, which are QTLs (Maxson & Canastar 2003, Nadeau 2003). Experiments aimed at mapping these modifier QTLs (a forward genetic experiment) are critical for achieving a full understanding of the genetic pathways and gene-gene interactions that affect complex traits.

Challenges in identifying aggression QTLs

Excluding reverse genetics studies (Miczek et al 2001), very few chromosomal regions have been identified that affect aggressive behaviours in mice, and even fewer have been mapped as part of a whole genome scan. This scarcity of mapped aggression QTLs is likely due, at least in part, to the following unusual challenges that aggressive behaviours pose as phenotypes for forward genetic studies:

- the complexity of aggressive social interactions, which are affected by many non-genetic (environmental and random) factors and gene-environment interactions;
- uncertainty about the best way to measure and quantify the many endophenotypes related to aggression in genetic studies; and
- the labour-intensive methods required to accurately measure this complex phenotype in hundreds of genetically-unique mice (e.g. intercross mice).

Non-genetic factors that affect aggressive behaviours

To date, most mouse behaviour QTL studies have focused on behaviours of individual mice, each tested alone in its home cage or in a testing apparatus (e.g. anxiety-related behaviours of an individual mouse in an open field; Flint 2003). In contrast, aggressive and other social behaviours occur, by definition, as part of an interaction between at least two mice. Therefore, aggressive behaviours are more complex phenotypes that are likely affected by a larger number of non-genetic factors, and are therefore more challenging to dissect using forward genetic methods. These non-genetic factors have less of a confounding effect on reverse genetics experiments, in which one measures the phenotype of many mice of the same genotype; in contrast, these factors may have a major impact on QTL mapping experiments, in which there is usually only a single mouse of each genotype (e.g. in an intercross population). There are, however, steps that can be taken to control these non-genetic factors in QTL mapping experiments, as outlined below.

Aggression is usually measured in a 'resident–intruder' test by observing the behaviour of a mouse of interest (the 'resident' mouse) towards a second ('opponent' or 'intruder') mouse that is placed in the resident's home cage. Alternatively, aggressive interactions can be observed between pairs of mice in a 'neutral' arena (Miczek et al 2001, Roubertoux et al 1999). In QTL mapping studies and other forward genetic studies, the aggressive behaviours of hundreds of resident mice, each of which has a unique genotype, must be scored and compared. However, variability in the degree to which the opponent (the intruder) tends to provoke or inhibit an aggressive response may be a confounding factor in comparing the aggressive phenotypes of resident mice. For example, the genotype of the opponent mouse can affect the aggression level of the resident mouse (Maxson & Canastar 2003). Therefore, it is advisable to use mice from a single inbred strain as a standard opponent in an aggression QTL study, which will reduce, but not eliminate, the variability among opponents' behaviour (Roubertoux & Carlier 1987). If one is interested in the offensive aggression of a resident mouse, it is best to pick standard opponents that are unlikely to initiate aggressive behaviours, such as a very unaggressive inbred mouse strain (e.g. A/J mice) or a partially restrained opponent.

Other non-genetic factors that can affect a mouse's propensity for aggressive behaviours include experiences with its mother or father (maternal or paternal effects), postweaning housing conditions, the manner in which mice are handled by testers, the type of aggression test used (resident–intruder vs. neutral cage) and random factors (e.g. noise in the testing area) (Guillot et al 1995, Maxson & Canastar 2003). Also, aggressive behaviours may be increased or decreased by particular pharmacological agents (Miczek et al 2002).

Postweaning housing conditions may serve as a useful example of these important environmental factors. If male mice are housed with other males after weaning, dominance hierarchies are formed through fighting and threat behaviours, and variability among mice in these social experiences can affect their later propensity for aggressive behaviours (Blanchard et al 2003, Maxson et al 2001). By socially isolating (individually housing) mice for a period of weeks prior to aggression testing, one can minimize the effect of recent fighting experience on aggressive behaviours. Depending on the inbred strain, social isolation (individual housing) may heighten or dampen aggressive behaviours. A period of social isolation is now a standard protocol prior to aggression testing (Maxson & Canastar 2003, Miczek et al 2001). Moreover, it is generally advisable to repeatedly test each mouse for aggressive behaviours, in order to get a more accurate measure than a single test can provide (Miczek et al 2001). The aggression level of a mouse may increase, decrease, or remain unchanged over repeated tests, depending partly on the genotype of the mouse (Miczek et al 2001, Ogawa et al 1998).

While stringent control of non-genetic factors will likely be necessary to make initial headway in mapping aggression QTLs, it is important to remember that such controlled conditions, such as social isolation, are artificial, and that the effect of any identified QTLs may be different under different environmental conditions. Ultimately, a full understanding of the genetics of aggressive behaviour will require a detailed understanding of these gene-environment interactions.

Ways of quantifying the endophenotypes of aggressive behaviours

The term 'agonistic behaviour' comprises the repertoire of different types of aggressive behaviours and reactions to them, including offensive behaviours, such as the attack bite, sideways threat and tail rattle (a threat behaviour), as well as defensive behaviours, such as the defensive upright posture and defeat posture. Different investigators have focused on different aspects of these various behaviours, including the latency until onset of the behaviour, or the number of the behaviours (e.g. number of tail rattles) in a given time period. The attack bite is generally the phenotype of greatest interest in genetic studies, because it is most intense and unambiguous manifestation of offensive aggression in mice (Miczek et al 2001). Composite scores of various aggression-related behaviours are potentially problematic if they equally weigh behaviours of different intensities (e.g. attack bites and tail rattles). The measurement of these aggressive behaviours is difficult to automate, i.e. generally requires human observation. The lack of automation, together with the need for repeated measurement, makes QTL studies of aggressive behaviours very labour-intensive.

Progress towards identifying QTLs that affect aggressive behaviours in mice

An example of aggression QTLs identified as part of a whole genome scan

One of the few studies to identify intermale aggression QTLs as part of a whole genome scan was published recently (Brodkin et al 2002). This study used NZB/B1NJ (extremely aggressive) and A/J (extremely unaggressive) inbred mice as parental strains. The methods chosen for housing and aggression testing were designed to control the effect of non-genetic factors on the phenotype. For example, male pups were individually housed immediately after weaning. A variant of the resident–intruder test was used in which the intruder was partially restrained, in order to minimize its ability to initiate aggressive behaviour. Intruders were all males from a less aggressive inbred strain, 129T2/SvEmsJ. The study focused on the attack bite as a relevant phenotype. Each mouse was tested on three consecutive days, using three different 129T2/SvEmsJ males as opponents and aggression level was scored as the number of tests in which an attack bite occurred. Each test was stopped immediately after the first attack bite (or at the end of 5 minutes, whichever came first), in order to prevent injury to the intruder. The phenotype of F1 hybrids was intermediate between the two inbred strains, suggesting a semi-dominant effect of the QTLs affecting aggression. Using a backcross breeding strategy, two QTLs were coarsely mapped, one on distal chromosome 10, and another on proximal chromosome X. Studies are now underway to accomplish the fine mapping of these loci. Possible candidate genes are listed in Table 1.

Other aggression QTLs

Several lines of mice have been selectively bred for high or low levels of offensive aggression, which confirms that a propensity for aggressive behaviours is partially heritable. These lines include the Turku aggressive (TA) and non-aggressive (TNS) strains bred in Finland, the NC900 and NC100 strains bred in North Carolina, and the short attack latency (SAL) and long attack latency (LAL) strains bred in the Netherlands (Miczek et al 2001).

In wild mice, there is evidence for a QTL affecting aggressive behaviours in a region of chromosome 17, the *t* region. The two major haplotypes observed in this region of chromosome 17 are wild-type (+/+) and the heterozygous *t* haplotype (+/*t*)—the *t*/*t* homozygote haplotype is lethal. A haplotype consists of a specific pattern of consecutive polymorphisms in a particular chromosomal location. The *t* haplotype is distinguished from wild-type by a series of inversion polymorphisms (Hammer et al 1991). Mice with the +/*t* haplotype showed greater offensive aggression than mice with the +/+ haplotype (Lennington 1991). This difference

TABLE 1 Chromosomal regions (coarsely mapped QTLs) that affect aggressive behaviours in mice

Mouse aggression QTLs	*Location (approximate)*	*Candidate gene(s)*	*Genetic background*	*Reference*
Chromosome 10	70 cM	*Dagk1*, *Avpr1a*	NZB/B1NJ and A/J	(Brodkin et al 2002)
Chromosome X	16 cM	*Gria3*, *Hprt*	NZB/B1NJ and A/J	(Brodkin et al 2002)
Chromosome 17 (t region)	20 cM	many	Wild mice	(Lennington 1991)
Y chromosome (non-pseudoautosomal)	3 cM	*Sry*	Studied in many backgrounds	(Maxson 1996, 2000)
X and Y chromosomes (pseudoautosomal region of Y chromosome)	75 cM (chromosome X)	*Sts*	NZB and CBA/H; SAL and LAL selectively bred strains	(Roubertoux et al 1994, Sluyter et al 1994)

cM, centiMorgan (a unit of genetic distance along a chromosome).
Dagk1 diacylglyerol kinase α subunit gene.
Avpr1a arginine vasopressin receptor 1A gene.
Gria3 glutamate receptor subunit AMPA3 gene.
Hprt hypoxanthine guanine phosphoribosyl transferase gene.
Sry sex determining region of chromosome Y.
Sts steroid sulfatase.

was found in the genetic background of wild mice, but not in the genetic background of laboratory mice (Miczek et al 2001).

Studies of reciprocal F1 hybrids, reciprocal backcrosses and reciprocal congenic strains, using various pairs of inbred progenitor strains, have suggested an effect of both the male-specific (non-pseudoautosomal, non-pairing) and the recombining (pseudoautosomal, pairing) part of the Y chromosome on intermale aggressive behaviours. *Sry* (the sex determining region) is a candidate in the non-pseudoautosomal region (Maxson 1996). The effect of this region on offensive aggression is modified by genetic background and maternal effects. The steroid sulfatase gene (*Sts*) is a candidate gene in the pseudoautosomal region of the Y chromosome (Le Roy et al 1999, Sluyter et al 1994). In addition, a spontaneous mutation in the androgen receptor gene (*Ar*) on the X chromosome (on a C57BL/6 genetic background), which greatly reduces expression of the gene, led to decreased aggressive behaviours (Maxson 2000) (see Table 1).

It is important to point out that the various aggression QTL studies mentioned above used differing pairs of parental inbred strains, differing postweaning housing environments, and differing aggression testing procedures. This is also true in reverse genetic studies of mouse aggression, where a variety of methodologies have been used (Miczek et al 2001). In order to make valid comparisons of results across studies, it will be important to attempt to standardize methods (Maxson & Canastar 2003).

Newer mouse genetics resources for mapping aggression QTLs

As mentioned above, a major impediment to forward genetic studies (e.g. QTL studies) of aggression has been the combination of two factors: the need to obtain an accurate measure of aggression in individual, genetically unique mice (e.g. intercross mice), on the one hand, together with the many non-genetic factors that can confound the measurement, on the other hand. A genetically segregating population with multiple mice of the same genotype would make it much easier to obtain an accurate phenotype for each genotype. Recombinant inbred (RI) strains are one such resource. RI strains are bred from the F2 progeny of two inbred strains, by 20 subsequent generations of brother-sister mating. Thus, each RI strain is fully inbred, is homozygous at every locus, and has a unique complement of genetic material from the progenitor strains (Brodkin & Nestler 1998). In the past, the data sets of RI mouse strains have been too small to provide sufficient statistical power for accurate QTL mapping (Brodkin & Nestler 1998, Silver 1995). However, the data sets of RI strains are now being greatly increased, which will likely increase their usefulness for QTL mapping (Peirce et al 2004). In addition, the recent creation of chromosome substitution (i.e. consomic) strains, will provide another such resource with multiple mice available for each genotype (Singer et al 2004). Each chromosome substitution strain has a chromosome from one inbred strain (e.g. chromosome 1 from C57BL/6J) on the genetic background of a second inbred strain (e.g. chromosomes 2–19, X, and Y from A/J). The availability of full sets of these chromosome substitution strain sets will facilitate the mapping of aggression QTLs. Finally, online databases of gene expression in various mouse brain regions are now available for C57BL/6J (B) and DBA/2J (D) inbred mice, their F1 hybrids, and a set of RI strains derived from B and D strains, which will facilitate the testing of candidate quantitative trait genes in these common inbred strains (Chesler et al 2004).

Summary

Quantitative trait locus analysis is a forward genetic strategy that provides a very important complement to reverse genetic strategies, such as the analysis of induced mutants (e.g. knockouts and transgenics). In order to identify modifiers of mutations, as well as alleles underlying natural variation in aggressive behaviours, QTL analysis will be necessary. However, progress in identifying aggression QTLs has been slow, perhaps because of the complexity of aggressive behaviours. Nevertheless, a number of aggression QTLs have been coarsely mapped, and are currently being fine mapped. New mouse genetic resources, such as expanded recombinant inbred strains and chromosome substitution strains, may facilitate future QTL analysis of aggressive behaviours.

Acknowledgements

Edward S. Brodkin is a recipient of a Burroughs Wellcome Fund Career Award in the Biomedical Sciences. This investigation was also supported by National Institutes of Health, Mentored Clinical Scientist Development Award KO8-MH068586-01 and by a NARSAD Essel Foundation Award. I thank Geena Mary V. Sankoorikal for assistance in preparing the figures, and Professors Wade Berrettini, Maja Bucan, and Thomas N. Ferraro for their comments on the manuscript.

References

Abiola O, Angel JM, Avner P, Bachmanov AA, Belknap JK, Bennett B et al 2003 The nature and identification of quantitative trait loci: a community's view. Nat Rev Genet 4:911–916

Belknap JK, Hitzemann R, Crabbe JC, Phillips TJ, Buck KJ, Williams RW 2001 QTL analysis and genomewide mutagenesis in mice: complementary genetic approaches to the dissection of complex traits. Behav Genet 31:5–15

Blanchard RJ, Wall PM, Blanchard DC 2003 Problems in the study of rodent aggression. Horm Behav 44:161–170

Brodkin ES, Nestler EJ 1998 Quantitative trait locus analysis: a new tool for psychiatric genetics. Neuroscientist 4:317–323

Brodkin ES, Goforth SA, Keene AH, Fossella JA, Silver LM 2002 Identification of quantitative trait loci that affect aggressive behavior in mice. J Neurosci 22:1165–1170

Chesler EJ, Lu L, Wang J, Williams RW, Manly KF 2004 WebQTL: rapid exploratory analysis of gene expression and genetic networks for brain and behavior. Nat Neurosci 7:485–486

Darvasi A 1997 Interval-specific congenic strains (ISCS): an experimental design for mapping a QTL into a 1-centimorgan interval. Mamm Genome 8:163–167

Darvasi A 1998 Experimental strategies for the genetic dissection of complex traits in animal models. Nat Genet 18:19–24

Doerge RW, Churchill GA 1996 Permutation tests for multiple loci affecting a quantitative character. Genetics 142:285–294

Flint J 2003 Analysis of quantitative trait loci that influence animal behavior. J Neurobiol 54: 46–77

Glazier AM, Nadeau JH, Aitman TJ 2002 Finding genes that underlie complex traits. Science 298:2345–2349

Guillot P, Carlier M, Roubertoux PL 1995 The Y chromosome effect on intermale aggression in mice depends on the test situation and on the recorded variables. Behav Genet 25:51–59

Hammer MF, Bliss S, Silver LM 1991 Genetic exchange across a paracentric inversion of the mouse t complex. Genetics 128:799–812

Hitzemann R, Malmanger B, Reed C, Lawler M, Hitzemann B, Coulombe S et al 2003 A strategy for the integration of QTL, gene expression, and sequence analyses. Mamm Genome 14: 733–747

Lander E, Kruglyak L 1995 Genetic dissection of complex traits: guidelines for interpreting and reporting linkage results. Nat Genet 11:241–247

Le Roy I, Mortaud S, Tordjman S et al 1999 Genetic correlation between expression of the steroid sulfatase gene, mapped on the pairing region of the Y-chromosome, and initiation of attack behavior. Behav Genet 29:131–136
Lennington S 1991 The t complex: a story of genes, behavior and populations. Adv Study Behav 20:51–86
Manly KF, Olson JM 1999 Overview of QTL mapping software and introduction to map manager QT. Mamm Genome 10:327–334
Maxson SC 1996 Searching for candidate genes with effects on an agonistic behavior, offense, in mice. Behav Genet 26:471–475
Maxson SC 2000 Genetic influences on aggressive behavior. In: Pfaff DW, Berrettini WH, Joh TH, Maxson SC (eds) Genetic influences on neural and behavioral functions. CRC Press, Boca Raton, p 405–416
Maxson SC, Canastar A 2003 Conceptual and methodological issues in the genetics of mouse agonistic behavior. Horm Behav 44:258–262
Maxson SC, Roubertoux PL, Guillot P, Goldman D 2001 The genetics of aggression: from mice to humans. In: Martinez M (ed) Prevention and control of aggression and the impact on its victims. Kluwer Academic, New York, p 71–81
Miczek KA, Maxson SC, Fish EW, Faccidomo S 2001 Aggressive behavioral phenotypes in mice. Behav Brain Res 125:167–181
Miczek KA, Fish EW, de Bold JF, de Almeida RMM 2002 Social and neural determinants of aggressive behavior: pharmacotherapeutic targets at serotonin, dopamine, and γ-aminobutyric acid systems. Psychopharmacology 163:434–458
Nadeau JH 2003 Modifying the message. Science 301:927–928
Ogawa S, Washburn TF, Taylor J, Lubahn DB, Korach KS, Pfaff DW 1998 Modifications of testosterone-dependent behaviors by estrogen receptor-α gene disruption in male mice. Endocrinology 139:5058–5069
Peirce JL, Lu L, Gu J, Silver LM, Williams RW 2004 A new set of BXD recombinant inbred lines from advanced intercross populations in mice. BMC Genet 5:7
Roubertoux PL, Carlier M 1987 Difference between CBA/H and NZB mice on intermale aggression. Maternal effects. Behav Genet 2:175–184
Roubertoux PL, Carlier M, Degrelle H, Haas-Dupertuis MC, Phillips J, Moutier R 1994 Co-segregation of the pseudoautosomal region of the Y chromosome with aggression in mice. Genetics 135:254–263
Roubertoux PL, Le Roy I, Mortaud S, Perez-Diaz F, Tordjman S 1999 Measuring aggression in the mouse. In: Crusio WE, Gerlai RT (eds) Handbook of molecular-genetic techniques for brain and behavior research (techniques in the behavioral and neural sciences). Elsevier, Amsterdam, p 696–709
Silver LM 1995 Mouse genetics: concepts and applications. Oxford University Press, Oxford, UK
Singer JB, Hill AE, Burrage LC et al 2004 Genetic dissection of complex traits with chromosome substitution strains of mice. Science 304:445–448
Sluyter F, van Oortmerssen GA, Koolhaas JM 1994 Studies on wild house mice IV. Differential effects of the Y chromosome on intermale aggression. Aggress Behav 20:379–386
Su AI, Wiltshire T, Batalov S, Lapp H, Ching KA, Block D et al 2004 A gene atlas of the mouse and human protein-encoding transcriptomes. Proc Natl Acad Sci USA 101:6062–6067

DISCUSSION

Martinez: You say that one of the limitations of this research is the standardization of the methods. I don't think this is a problem. We have worked on mice for many years. With regards to the opponent, there are many possibilities of getting a standard opponent. One of the methods we use is to make the intruder anosmic.

Brodkin: Certainly, there are various methods now available for standardizing opponents. You could use intruders from the same inbred strain, for example. I was referring not only to standardizing intruders, but also to many other subtleties of the way that the aggression testing is designed. I don't mean to suggest that everyone must follow only one standardized protocol in studying aggressive behaviours. This would limit science. But I am suggesting that subtle differences in the ways that studies are carried out can make it difficult to compare results across labs. We discussed earlier the differences between results of the Maxson group and the Roubertoux group. Maxson & Canastar (2003) have recently reviewed the effects of seemingly subtle methodological differences on genetic studies of aggressive behaviours.

Martinez: What do you think accounts for those differences? Were they looking at different behaviours?

Brodkin: Maxson & Canastar (2003) identified several important methodological factors that can have an impact on results, including the maternal environment, the type of opponent/intruder, postweaning housing and the setting of the test, i.e. resident home cage vs. neutral arena. Depending on how you carry out the aggression test you may be tapping into different aspects of aggressive behaviour, so different chromosomal regions might be affecting these different aspects of the phenotype.

Craig: In your genome scanning, what were the effect sizes for the peaks on chromosomes 10 and X for the total variability?

Brodkin: In our QTL mapping study, we looked at aggression as a threshold trait. We looked at which mice had reproducible aggressive behaviour, showing an attack in at least two out of three tests. We divided our population in a qualitative way into those who showed reproducible aggression and those who didn't. Because we analysed the data in this way, we couldn't calculate variance or the size effect of each QTL. I'm now in the process of re-analysing the data using a quantitative scale in order to look at the effect sizes of each QTL. Also, once we have completed breeding congenic strains we will hopefully be able to see the effect size of each locus.

Craig: I'd like to ask an unrelated question to do with the MAOA locus on the X. Is anything known about the activity levels in the two parental strains? Do they differ?

Brodkin: Not that I know of. We are sequencing MAOA in the two parental strains and we haven't found any coding region polymorphisms to date. The exon sequences are the same. Now we are trying to look upstream at some of the promoter regions. There is this well known polymorphism in humans that controls expression levels of MAOA. I know of no similar polymorphism in mice.

Lesch: There is none. It is specific to primates.

Dulac: To be a bit provocative, what if the QTL approach isn't the right method to investigate this sort of problem? Let's imagine that we have 20 or 30 genes involved. It is clear that there are strains of mice that are extremely different in their aggressive behaviour, but let's say that there are 20 genes, each providing a moderate increase in aggression. How will you be able to study this? It will be too complicated, because each chromosome will carry many alleles. Do you have any way to estimate the number of genes that might be involved in the traits?

Brodkin: Your question gets at the important and controversial issue of the relative value of the QTL approach, compared to other forward and reverse genetic approaches, in dissecting the genetics of complex traits (see Nadeau & Frankel 2000, Belknap et al 2001). As you point out, it is very challenging to identify the relevant genes in QTL regions because there are multiple QTLs and genes that affect a complex trait like aggression, and therefore the task is very complex compared with identifying a single mutation of large effect on a phenotype. There are statistical methods that can be used prior to a mapping experiment to estimate the number of genes that affect a quantitative trait, such as Wright's polygene estimate. However, these estimates rely on a number of simplifying assumptions that may or may not prove correct when the loci are mapped empirically (e.g. the assumption that each of the genes makes an equivalent contribution to the phenotype) (Silver 1995). Nevertheless, despite the complexities, people have remained interested in QTL analysis as a strategy for discovering naturally occurring gene variants that account for phenotypic variation. After coarsely mapping QTLs, one way of simplifying the genetic analysis and maximizing chances for success in identifying genes underlying QTLs is to develop a congenic strain in which a QTL region from one inbred strain is transferred to the genetic background of the other inbred strain. This isolates that one QTL region for further analysis and reduces the number of genes segregating in any further fine mapping crosses. This strategy has been successful in identifying genes underlying QTLs in animals and plants (see for example Shirley et al 2004, Korstanje & Paigen 2002, Fridman et al 2004). I should also point out, though, that even in a single QTL region isolated in a congenic strain, it is possible that there is more than one allele that affects the phenotype. So, you have a fair point about the challenges and complexities of QTL analysis.

Koolhaas: There are different questions underlying both approaches. The QTL approach addresses the question of where the genetic variation is. This is a completely different question from whether a gene is involved or not. There are many genes involved in the neurobiology of aggression, but QTL is looking just for those responsible for variation.

Nelson: You could find some olfactory gene that could be involved in this.

Koolhaas: Of course. Olfaction is involved in aggression and, for that reason, all genes involved in the physiological processes underlying olfaction are by definition involved in aggression.

Dulac: The problem is that if you have hundreds of genes involved, are you going to identify any?

Koolhaas: The molecular mechanisms obviously include hundreds of genes. However, it is unknown so far whether all of these genes may be polymorphic and hence contribute to the variation in aggression. Alternatively, we may expect genetic polymorphism on a limited number of genes coding for proteins that play a key role in the neurobiology of aggression.

Craig: Perhaps we should move in the direction of association studies, collecting very large samples from the wild. You can detect effect sizes of 1%. It would be a huge project.

R *Blanchard:* One of the basic problems is that people think that aggression is simple and genes are complex. I think aggression is actually extremely complex. Here, I am a little concerned: the primary dependent variable is the latency of attack. Yet Jap's work would lead us to believe that attack latency is a measure more of coping and stress-related behaviour than aggression.

Koolhaas: It starts with what you want to explain. If you see a correlation between aggression and a wide variety of other behaviours, then the question is what do you want to explain? The 'umbrella' or behavioural syndrome, or attack latency as a simple measure, or biting?

Brodkin: In this particular study, I was interested in the attack bite as the primary phenotype.

R *Blanchard:* It is the latency to attack bite, rather than the follow-through with a consistently motivated aggression behaviour.

Brodkin: I stopped the attack after the initial bite or two in order to minimize injury to the partially-restrained intruder mouse. I don't know whether we have adequately addressed your question.

Manuck: I am also a little concerned about the power of QTL linkage. The circumstance you described was one in which you had about 100 markers with a 20–30 cM separation, and perhaps 300 animals. Let's say that the phenotypic heritability of the trait you wish to study is about 0.5 and that there are 20 genes containing sequence variation contributing to these individual differences. For the sake of argument, let's say that variation in each of these several genes contributes equally, thus each accounting for a very small portion of inter-individual variability in the phenotype. Would coarse mapping with these parameters have the power to detect QTL heritabilities of this magnitude?

Brodkin: Such a course mapping study with only about 400 mice would be unlikely to detect a QTL that accounts for only 2.5% of the phenotypic variance, but it should detect a QTL that accounts for approximately 10% of the variance (Lynch & Walsh 1998, Darvasi 1998). QTLs of this magnitude of effect on neurobiological or behavioural traits have been found fairly commonly in crosses between inbred mouse strains (see e.g. Wehner et al 1997). The general rule is that the smaller the effect size of a QTL on a phenotype, the larger the mapping population required to detect it. Keep in mind that, although there are likely to be very large numbers of genes and alleles that affect aggressive behaviour in an outbred population, only a smaller subset of these genes will be polymorphic when two inbred strains are compared. Therefore, in a breeding cross between two inbred progenitor strains as part of a QTL mapping project, there will be fewer polymorphic alleles of larger effect on the phenotype than there would be in an outbred population. It has been pretty straightforward to coarsely map QTLs that affect behavioural differences between pairs of inbred strains. It has been much more difficult to get from coarsely mapped QTLs to the relevant genes in those QTL regions. Studying knockouts may be addressing a different question from QTL mapping. Knockouts provide information about the effect of obliterating expression of a particular gene, whereas QTL analysis is aimed at identifying the more subtle, quantitative effects of naturally-occurring alleles on a phenotype of interest.

Dulac: I am not talking just about knockouts, but ENU mutagenesis.

Brodkin: ENU mutagenesis is, of course, a great approach for dissecting the genetics of complex traits, and I view ENU and QTL analysis as complementary approaches. As you know, an ENU-induced mutation is simpler (but certainly not easy) to identify, because the task is to identify a single mutation of large effect on the phenotype. However, QTL mapping may identify a different set of loci from those identified to date in ENU mutagenesis programs. Take, for example, the study of circadian behaviours in mice. Joseph Takahashi at Northwestern University identified the *clock* gene using ENU mutagenesis (King et al 1997). His group has also used QTL analysis to study circadian rhythms, and most of the QTLs he detected map to locations different from genes identified by mutagenesis (Shimomura et al 2001). You could argue that if you did a very large number of ENU mutagenesis studies, eventually you would come up with every gene that affected a phenotype in a mouse. But part of the goal of QTL mapping is to identify spontaneously occurring allelic variants in genes, so it is a somewhat different goal. It is often less efficient some of these other strategies, but it gets at different questions.

Skuse: You quoted Jonathon Flint's review paper (Flint & Mott 2001, Flint 2004). I remember having a discussion with him a few years back when he was looking for a QTL for anxiety in mouse. He recommended that, for certain purposes, it was better to use outbred strains in preference to inbred ones (Valdar et al 2003). What would the advantages of this approach be?

Brodkin: Flint's group has used genetically heterogeneous stocks of mice, derived from eight known inbred mouse progenitor strains, to achieve the fine mapping to less than 1 cM of a QTL that affects an anxiety-related behaviour in mice (Talbot et al 1999). Once a QTL has been coarsely mapped, these heterogeneous stocks are useful for fine mapping because they have many more recombinations along each chromosome, because they have undergone approximately 60 generations of intercrossing. Flint has pointed out that these heterogeneous stocks could also be used to map modifier QTLs that affect transgenes or knockouts (Flint & Mott 2001).

Lovell-Badge: He also claims that he can get 5% differences.

Brembs: With regard to the methodology of QTLs, in principle it is good to have convergent redundant information. Even though we may only be able to do a coarse mapping of QTLs, in conjunction with other methodologies this is always providing more reliability to the data. I wouldn't write QTLs off just yet.

Craig: We should remember that we are talking about QTL mapping by linkage, as opposed to QTL mapping by association approaches, which can be very powerful. I still have my doubts about QTL mapping by linkage.

Brodkin: I should also mention that my lab has done a second QTL mapping study using BALB/cJ mice as the aggressive strain and A/J as the unaggressive strain. One of the loci that we have mapped is in the same region mapped in our original cross of NZB/B1NJ and A/J mice — the telomeric region of chromosome 10. Our hypothesis is that there may be an A/J allele(s) in this region of chromosome 10 that reduces tendency to initiate aggressive attacks (Brodkin et al 2004).

Brembs: QTL analysis can also address the question of necessity and sufficiency in a different way than knockouts would. Can you supplement QTL analysis in the fine mapping procedure by methods such as representational difference analysis or gene-microarrays? Would they help you? They have their own flaws, but in combination with QTLs a lot of false positives would just fall out. Are you considering supplementing QTL studies with more molecular approaches such as these?

Brodkin: Yes we are. The question is which brain regions to look at.

Pfaff: With microarrays you do best if you have an experimental design using an agent that can drive the behaviour as a function of time.

Brodkin: I wanted to add another point regarding the analysis of mutant mice. When you knock out a gene or mutate a gene in some way, you may see an effect on a phenotype of interest, such as aggressive behaviour. However, if you then transfer that knockout to a different genetic background (e.g. from a mixed 129 substrain/C57BL/6 background to a C57BL/6 inbred strain), you often see a different effect on phenotype because of the effect of modifier loci. These modifiers are, in essence, QTLs. Therefore, QTLs are of importance even to those primarily interested in the analysis of single gene mutations.

Dulac: The challenge is to choose the right method to find those genes, not whether or not there are modifiers to be found. In fact, I think there are too many modifiers for aggressive behaviour which makes it very difficult.

Brodkin: In a sense the modifiers are annoying because you can lose your phenotype when transferring a mutation to a new genetic background. But in another sense, the modifiers are very interesting because they are somehow part of the same pathway, modifying the effect of the mutated gene. I agree, however, that if there are many modifiers, they may be very challenging to identify.

C Blanchard: I would like to put a couple of things together. It seems that you have gone to great lengths to try to standardize your test situation. I wonder if under the circumstances that aggression is so responsive to so many other things, it might be a good idea to pay a little attention to standardizing the test animal before you start. We know well that fear interacts magnificently with aggression. You could probably get effects on aggression through fear more easily than through altering aggression. If you are going to compare strains X and Y and they are differentially aggressive, why can't you make sure before you start that these are not differentially aggressive because they are differentially fearful, or because they are differentially able to smell other male pheromonal products, in an effort to reduce false positives due to these differences that are extraneous to the genes that underlie the aggressive motivation?

Brodkin: So you're saying that fear would be extraneous to what underlies aggressive motivation itself?

C Blanchard: Yes, fear can act on aggressive motivations, but then it can act on everything else. If you put a rat in with a cat and give him a lady rat he's not going to show any sexual behaviour because of his fear. It is an extraneous but very powerful suppressive effect. Of course, some animals are more fearful or anxious than others.

Brodkin: So you are suggesting that, prior to embarking on a QTL mapping project, one should try to understand, to the greatest extent possible, the reasons for the observed phenotype differences between the inbred strains. I think that is very important. For example, prior to embarking on our QTL analyses of aggressive behaviour, we asked whether A/J mice, which show such a low level of aggression, are actually capable of aggressive attack behaviour at all. Although A/J mice showed extremely low aggressive attack behaviour in the context of our testing procedure, A/J mice do show a low level of aggressive behaviour in their home cages when a new male is introduced, which implies that A/J mice are not completely incapable of showing aggressive behaviour due to some sort of motor problem or other issues extraneous to aggressive motivation. But I think it is a great idea, as you are suggesting, to more fully characterize some of these other phenotypes.

Pfaff: Your chromosome 10 locus has vasopressin 1A receptor in it. Do you predict that signalling through the VP1A receptor would increase or decrease aggression?

Brodkin: It is hard to predict that using my data alone. If *Vp1a* were the relevant gene in that locus, there would be some allelic difference in the gene or the gene regulatory regions between NZB and AJ, and it could be that the NZB allele tends to heighten aggression or that the AJ allele tends to lower it, for example. From Craig Ferris' data, I would guess that the increased signalling through the VP1A receptor would tend to increase aggression. So, if *Vp1a* turned out to be the relevant gene, I would guess that NZB mice would show higher levels of expression of *Vp1a* or higher signalling through VP1A in particular brain regions, which would tend to heighten aggressive attack behaviour in NZB mice.

Ferris: What are we gleaning from the forward genetic approach versus a reverse genetic approach? Everyone would agree that this is a complex individual problem of impulsivity and violence, and that it is regulated by many genes. Each of these different technologies is getting us closer to the cluster of candidate genes that are important. There are likely to be thousands, but you could put valence to the individual genes and say that one is more important than another. Understanding leads to intervention, and in the end we will have two intervention strategies based on knowledge of the genes: we will have drugs to stop impulsivity and violence, or because we understand how the environment interacts with these genes we will take a psychosocial approach. My feeling is that all of these approaches are excellent, and collectively will help to focus on the key players.

Olivier: In my lab we use both a reverse genetics and a forward genetics approach. There is the example of the serotonin $(5\text{-HT})_{1A}$ knockout mouse. This has been made by three independent groups on three different backgrounds, and all three were more anxious than wild-types. This fits in with the idea that 5-HT_{1A} receptor agonists are anxiolytic in humans and animals. As far as I know Jonathon Flint has not come up with a QTL for the 5-HT_{1A} receptor. We know its chromosomal location, and we also have chromosomal strains in Utrecht. We were interested to see whether on that particular chromosomal strain we could find something back from that anxiety. I would have expected it to come from different sites—that there would be a final common pathway. The problem was complicated. The reverse genetics and pharmacology said yes, and the forward genetics said I don't know.

Lovell-Badge: If it is not a natural variation on your two starting populations, then you won't find it. Another strategy is to use ENU mutagenesis as a sensitized screen, looking for enhancers or suppressors of an already-existing phenotype. People are doing this in the mouse now.

References

Belknap JK, Hitzemann R, Crabbe JC, Phillips TJ, Buck KJ, Williams RW 2001 QTL analysis and genomewide mutagenesis in mice: complementary genetic approaches to the dissection of complex traits. Behav Genet 31:5–15

Brodkin ES, Silver LM, Dow HC, Kaercher KA, Sankoorikal GMV 2004 Quantitative trait locus analysis of intermale aggressive behaviors in BALB/cJ and A/J mice: initial genome scan results. Society for Neuroscience 2004 Abstract Viewer, Society for Neuroscience, Washington DC (online)

Darvasi A 1998 Experimental strategies for the genetic dissection of complex traits in animal models. Nat Genet 18:19–24

Flint J 2004 The genetic basis of neuroticism. Neurosci Biobehav Rev 28:307–316

Flint J, Mott R 2001 Finding the molecular basis of quantitative traits: successes and pitfalls. Nat Rev Genet 2:437–445

Fridman E, Carrari F, Liu Y-S, Fernie AR, Zamir D 2004 Zooming in on a quantitative trait for tomato yield using interspecific introgressions. Science 305:1786–1789

King DP, Zhao Y, Sangoram AM et al 1997 Positional cloning of the mouse circadian clock gene. Cell 89:641–653

Korstanje R, Paigen B 2002 From QTL to gene: the harvest begins. Nat Genet 31:235–236

Lynch M, Walsh B 1998 Genetics and analysis of quantitative traits. Sinauer Associates, Sunderland, MA, p 431–484

Maxson SC, Canastar A 2003 Conceptual and methodological issues in the genetics of mouse agonistic behavior. Horm Behav 44:258–262

Nadeau JH, Frankel WN 2000 The roads from phenotypic variation to gene discovery: mutagenesis versus QTLs. Nat Genet 25:381–384

Shimomura K, Low-Zeddies SS, King DP et al 2001 Genome-wide epistatic interaction analysis reveals complex genetic determinants of circadian behavior in mice. Genome Res 11:959–980

Shirley RL, Walter NAR, Reilly MT, Fehr C, Buck KJ 2004 Mpdz is a quantitative trait gene for drug withdrawal seizures. Nat Neurosci 7:699–700

Silver LM 1995 Mouse genetics: concepts and applications. Oxford University Press, New York

Talbot CJ, Nicod A, Cherny SS, Fulker DW, Collins AC, Flint J 1999 High-resolution mapping of quantitative trait loci in outbred mice. Nat Genet 21:305–308

Valdar WS, Flint J, Mott R 2003 QTL fine-mapping with recombinant-inbred heterogeneous stocks and in vitro heterogeneous stocks. Mamm Genome 14:830–838

Wehner JM, Radcliffe RA, Rosmann ST et al 1997 Quantitative trait locus analysis of contextual fear conditioning in mice. Nat Genet 17:331–334

Genes for sex hormone receptors controlling mouse aggression

Donald Pfaff, Elena Choleris and Sonoko Ogawa

Laboratory of Neurobiology and Behaviour, The Rockefeller University, 1230 York Avenue, New York, NY 10021, USA

Abstract. The stereotypies of mouse social behaviours have encouraged their systematic analysis in this genetically tractable animal. Following experiments with genes for nuclear receptors and other neuroendocrine genes, we can state seven 'lessons' of gene/behaviour causal relations bearing on sociosexual and aggressive behaviours. The effect of a given gene on a given behaviour depends upon: (1) exactly when and where that gene is expressed in the brain; (2) the gender of the animal in which it is expressed; (3) the age of the animal; (4) the nature of the opponent; and (5) the form of aggression (e.g. testosterone-facilitated aggression vs. maternal aggression). (6) Better social recognition is correlated with lower levels of aggression. We have gathered evidence for a four-gene micronet involving oestrogen receptors α and β, oxytocin, and the oxytocin receptor as expressed in the hypothalamus and amygdala. (7) Some genetic influences on aggression derive from their effects on fundamental, generalized arousal of the mammalian brain, which underlies the expression of any emotional behaviour.

2005 Molecular mechanisms influencing aggressive behaviours. Wiley, Chichester (Novartis Foundation Symposium 268) p 78–95

Molecular mechanisms of aggression have tended to focus on two important causes well represented in this Novartis Foundation symposium. First comes the dominant role of serotonergic neurotransmission. Second comes the facilitation of aggression by androgenic hormones, especially testosterone. The latter, hormonal influences have received extensive review by Simon and his colleagues (Simon 2002), with respect to the developmental effects of hormones on the organization of aggression-related pathways in the brain, as well as the aggression-promoting actions of testosterone in the adult brain. In the adult, testosterone does not trigger aggression *in vacuo*, but potentiates aggressive responses to physical provocation (McGinnis et al 2002).

Considering the functional genomics of aggression in mice, in this paper we will argue three general points. First, the complexity of the mammalian CNS has forced a more sophisticated kind of thinking than used to be evident about how genes influence behaviours. Historically, functional genomic thinking was dominated

TABLE 1 Comparisons of ERα and ERβ functions in the CNS

Necessary?	*Sufficient?*		*Assays in the ♀*	*Assays in the ♂*
ERα and β	Neither α nor β	ERα and β must synergize	Social recognition	None
Neither α nor β	Either α or β	ERα and β can substitute for each other	E induction of PR (ICC) E reduction of ER (ICC)	E induction of PR (ICC) E reduction of ER (ICC) Simple mounting
α, β each for its own	α, β each for its own	ERα vs β contri-butions not related	Maternal behaviour (α) Suppression of aggression (α) Reduction of food intake (α) Reduction of anxiety (α)	Intromission and ejaculation Anxiety response (α)
ERα absent β	ERα absent β	ERβ can reduce α effect	Lordosis behaviour	Aggression
Optimal α/β balance	Optimal α/β balance	ERα vs β always opposed	None	None

Reference: chapter 47 in Pfaff et al (2002).

by the classical work of Beadle and Tatum. Using the fungus *Neurospora*, they discovered biochemical deficits following certain mutations which led to their central formulation: the 'one gene/one enzyme' formulation. For the mammalian CNS, things are more complicated. Working with null mutations for sex hormone receptors (Table 1, Chapter 47 in Pfaff et al 2002) we see that different combinations of hormone receptor gene activity permit certain behaviours associated with reproduction in both male and in female mice. That is, we have proved that different *patterns* of gene expression are required for different *patterns* of sociosexual behaviours.

Secondly, with well-controlled experimentation in mice we can discern how particular genetic influences on behaviour depend on other features of the animal and its environment. In a form which depends heavily on the article by Ogawa, Choleris and Pfaff (Ogawa et al 2005), these are listed in the next section in the form of seven 'lessons' of gene/behaviour causal connections. In the third and final section, we show how to envision the large number of genes and considerable diversity of causal routes for genes affecting mouse aggressive behaviours.

Seven principles of gene/behaviour relations applying to natural social behaviours in mice

(1) If the gene which codes for the classical oestrogen receptor (ER)α is knocked out, the female mouse's behaviour is masculinized. She behaves more like a male and, in fact, is treated like a male by other mice (Ogawa et al 1996). In such an experiment, the gene's product is absent throughout the body and throughout the life of the mouse. In dramatic contrast, if the ERα function is interrupted during the neonatal critical period by the microinjection of an antisense DNA oligomer directed against the translation start site of ERα mRNA, only in the hypothalamus, then the masculinizing influence of testosterone during the neonatal critical period is opposed (McCarthy et al 1993). Reasoning from the comparisons between these two sets of results, we infer that *the effect of a given gene on a given type of behaviour depends on where in the brain and exactly when that gene is expressed.*

(2) Deleting the ERα gene permanently in a male mouse abolishes aggression (Fig. 1) (Ogawa et al 1998a). However, in a female mouse, the same type of mutation increases aggression (Fig. 2) (Ogawa et al 1998b). The results for the two sexes are opposite. Now consider deletion of the gene coding for ERβ. In males, such a gene knockout can increase aggression (Nomura et al 2002). However, in female mice, knocking out ERβ reduces testosterone-facilitated

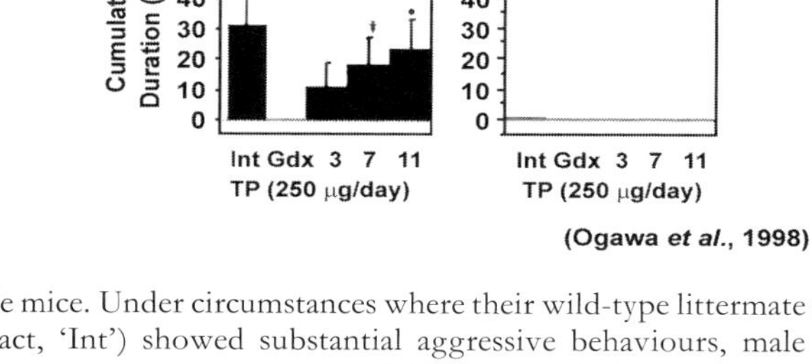

FIG. 1. Male mice. Under circumstances where their wild-type littermate controls (alpha WT, gonadally intact, 'Int') showed substantial aggressive behaviours, male mice with the null mutation for oestrogen receptor α (αERKO) had their aggressive behaviour virtually abolished. This was not simply because of inadequate testosterone levels, because castration (in which no mice showed aggression) followed by supplementation with testosterone propionate (TP) still did not yield any aggressive behaviour for the αERKO male mice. (From Ogawa et al 1998a with permission.)

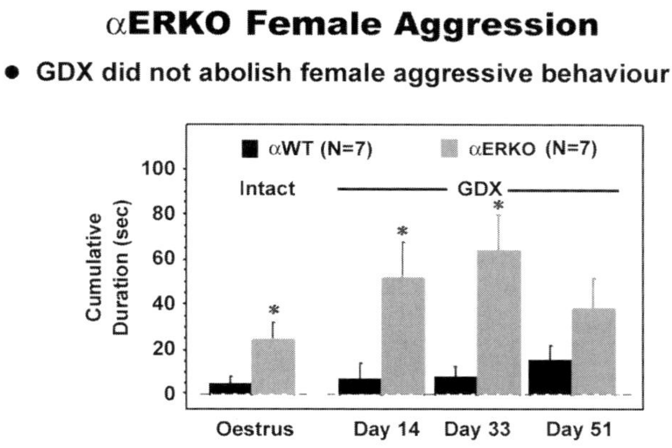

FIG. 2. Female mice. αERKO females showed more aggression than their wild-type (WT) littermate controls. This could not have been due to ovarian hormone differences because the behavioural effect of knocking out ERα persisted after gonadectomy (GDX). (From Ogawa et al 1998b with permission.)

aggression (Durbak et al 2002). From these two sets of contrasts between male and female mice, we infer that *the effect of a given gene on aggression depends on the gender in which that gene is expressed.*

(3) It is obvious that with respect to aggressive behaviours, the magnitude of the phenotype in the ERβ knockout male mouse declines with age (Nomura et al 2002). We do not know why. Further, when measuring locomotor behaviours in ERα knockout female mice compared to their littermate wild-type controls, we see that the phenotype (failure to respond to oestrogens with increased locomotion) gets stronger with age (Garey et al 2003). We do not know the mechanism of that age-related change either. Nevertheless, it is clear that *the effect of a specific gene on aggressive (or locomotor) behaviours can depend on the age of the animals at which the behavioural assay is conducted.*

(4) During the resident–intruder paradigm for testing aggression, ERα knockout female mice display high levels of increased aggression toward female intruders. Their aggression persists well beyond that shown by their wild-type littermate controls, whether or not the intruder is a female mouse treated with oestrogens and progestins. In contrast, if the intruder is an olfactory bulbectomized male mouse, the ERα knockout female's aggression is at a very low level, not any different from that of the wild-type female mouse. This comparison shows that *the effect of a specific gene on aggressive behaviour can depend on the nature of the opponent.*

A subtle variation upon this argument leads to a new point. Recent research with oxytocin knockout (OTKO) mice has shown that these null mutants produce odours that are discriminably different from those emitted by their wild-type littermate controls (Kavaliers et al 2004). As a result, in any social environment, the behaviours of other mice could be altered by the distinct odours they are receiving from OTKO mice. As a consequence of that fact, the resultant behaviours of the OTKO mice themselves could be changed. Thus, we have inferred that *some genetic effects on aggressive behaviours could be due to causal chains routed through the behaviours of social partners*.

(5) Female mice tested for their vigour in nest defence during preliminary experiments (Le et al 2002) showed a marked effect of the null deletion of ERβ. At the beginning of each test, ERβ knockouts showed a much greater number of attacks than wild-type controls. On the other hand, when tested for testosterone-facilitated aggression (Durbak et al 2002) ERβ knockouts responded with significantly less frequent aggression for a given dose of testosterone. This comparison suggests that *the effect of a specific gene on aggressive behaviour can depend on the type of aggression tested*.

(6) *Social recognition and bonding helps to reduce aggression*. In mice, social behaviour depends much more on odours and pheromones than is true for human beings. Such chemosensory information, travelling along traditional olfactory pathways from the main olfactory bulb or sensed by the vomeronasal organ and travelling along pheromonal pathways from the accessory olfactory bulb, converges on the amygdala. Here oxytocin acts to facilitate social recognition (Ferguson et al 2002). If a mouse recognizes the chemical signals from a conspecific, affiliative, friendly behaviours are encouraged and aggression is discouraged.

On the basis of a large body of new data demonstrating ERα knockout, ERβ knockout and OTKO effects on social recognition in mice, Choleris et al (2003a) have put together the theory of a four-gene micronet which supports this crucial behavioural function. Oestrogens working on gene transcription rates through ERβ in the hypothalamus and through ERα in the amygdala provide for efficient, high volume oxytocin signalling, thus to facilitate social recognition. A somewhat more precise approach to social recognition by animals, the Binary Choice assay, supports Choleris' published conclusions (Choleris et al 2003b). Further, the concept has been extended to an ecologically relevant paradigm in which the social recognition function can be related to avoidance of parasitic infection (Kavaliers et al 2003). If the Choleris theory is correct, then interruption of oxytocin receptor (OTR) production in the amygdala should reduce social recognition in the treated mice. As predicted, microinjection of antisense DNA oligomers in the form of locked nucleic acids housed in biodegradable microspheres (Choleris et al 2003c) significantly reduced social recognition in that—compared to control animals—the mice with antisense-loaded microspheres in their amygdalas were less able to recognize that a new intruder mouse had been introduced, rather than a mouse with which they should have become familiar. Since, even at this early stage in the functional genomics of this system, four genes operating in two regions of the basal forebrain are involved, we underline our reformulation of the classical Beadle and Tatum principle quoted above, to state that for mammalian brain mechanism, *patterns and combinations* of genes are required to support *patterns* of mammalian behaviours.

(7) *Genetic controls on elementary arousability of the CNS influence the expression of aggressive behaviours.* In order to reformulate the classical 'arousal problem' (Pfaff 2005), we had to devise a precise mathematical approach to the resolution of a false dichotomy historically present in this field, as follows: on the one hand, a great deal of electrophysiological evidence from recordings across the cerebral cortex following manipulations of the brainstem said 'yes', there is a monolithic generalized arousal function in the mammalian CNS. However, some cognitive neuroscientists argued that the concept of arousal has become hopelessly subdivided. A clear theoretical resolution can be found, for the first time (Garey et al 2003), in an equation which combines both generalized (Ag) and various specific forms of arousal (As_{1-n}) such as sex, hunger, fear, etc.

$$A = F(KgAg + Ks_1As_1 + Ks_2As_2 + Ks_3As_3 \ldots + Ks_nAs_n). \tag{1}$$

Then, for a concrete, experimental approach to the arousal problem, we set forth

- (*i*) a clear operational definition of generalised arousal, and (*ii*) a mathematical approach:
- (*i*) *Operational definition.* A more aroused animal is (A) more responsive to a wide variety of external stimuli spanning sensory modalities; and (B) is more motorically active; and (C) is more emotionally reactive. This definition yields easily gathered quantitative, physical measures of activity.
- (*ii*) Rather than falling prey to the false dichotomy, mentioned above, that 'Generalized Arousal comprises 100% (vs. 0%)' of arousal mechanisms, we used Principal Components Analysis (PCA). PCA mathematically separates and analyses the variations of behavioural responses during experiments employing many mice and many arousal-related assays. Essentially, PCA 'lets the mouse tell the experimenters' how its arousal functions are structured, in quantitative terms. Our results showed that of all arousal-related assays tested, the generalized arousal component accounted for about one-third of the data (Garey et al 2003).

In turn, we make the rather obvious point that the force of CNS arousal controls the amplitude and frequency of expression of any aggressive response in all mammals. Global arousal impacts aggression in different ways from its impact, for example on sleep, sex or fear (Table 2). So far, in our lab, detailed analyses

TABLE 2 Partial tabulation of how fluctuations in arousal theoretically impact instinctive, biologically regulated behaviours

Arousal parameter	*From sleep to wake*	*For sex, fear*	*For aggression*
Amplitude	Zero to W	X, Y	Z
Time	24 hour rhythm	'Now' (upon S)	'Now'
Specific stimulus?	No	Yes	Yes
A typical transmitter	Histamine	Norepinephrine	Serotonin
Neuroanatomy emphasized	TMN, POA	VMN, amygdala	Septum

POA, preoptic area; TMN, tuberomamillary nucleus; VMN, ventromedial nucleus.

have centred on three genes, those for ERα, ERβ and prostaglandin D synthetase (PGDS). Clearly, knocking out ERα reduced sensory alertness (Fig. 3, from Garey et al 2003) and reduced voluntary motor activity in the older animals (Fig. 4, from Garey et al 2003). The perfect genetic control — knockout of a likely gene duplication product, ERβ — had no significant effect on either of these measures. Thirdly, we followed up a result from a microarray experiment designed to look at oestrogen-sensitive genes. PGDS mRNA was significantly elevated in the medial basal hypothalamus, but was regulated in the opposite direction in the preoptic area (Mong et al 2003a).

Because certain preoptic area neurons crucially control locomotor activity and sleep, we followed up this finding. Antisense DNA oligomers microinjected in the preoptic area, directed against the mRNA coding for PGDS, caused the animals to be more active in terms of response to vestibular stimuli and in terms of voluntary motor activity (Mong et al 2003b). More generally, considering all of the neural pathways and neurochemical controls over arousal (Pfaff 2005), more than 50 genes are centrally involved in arousal neurobiology. Together they provide massive opportunities for explaining variations in an elementary neural function required for all cognitive and emotional behaviours. Because elevated levels of arousal are required for the visible expression of any emotional behaviour — and because highly over-aroused individuals may be susceptible to untoward displays of anger — genetic effects on elementary arousability provide interesting causal routes for explaining certain indirect genetic effects on aggressive behaviours.

Diversity of genetic influences and illustration of gene/environment interactions

From the point of view of developmental neuroendocrinology, genes on the Y chromosome leading the sexual differentiation of the brain toward a masculine phenotype would be most obviously involved and would help to explain sex

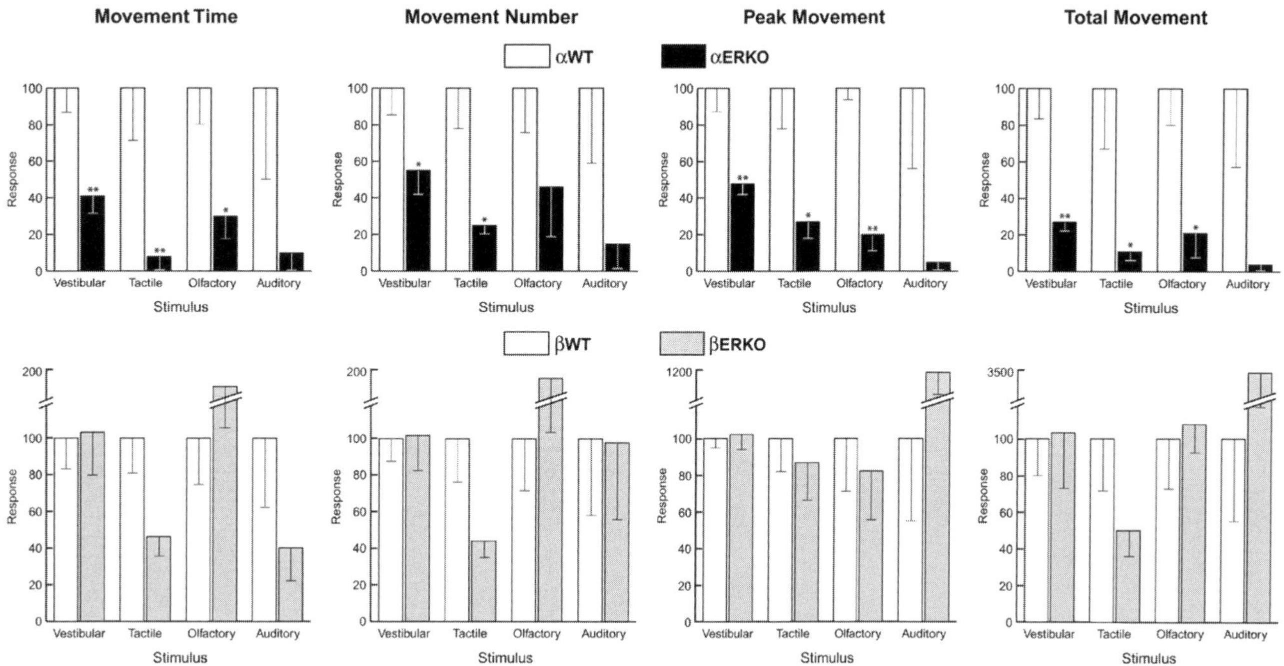

FIG. 3. Arousal, as measured by responsiveness to sensory stimuli. Expression of the gene for αER but not *β*ER is important for arousal in response to sensory stimuli in female mice. Tested in their home cages during the light parts of their daily cycles, female mice with the null mutation for the αER gene (αERKO) responded significantly less by several quantitative parameters to sensory stimuli in each of four stimulus modalities. There were no significant differences in arousal responses following knockout of a closely related gene (*β*ERKO). (From Garey et al 2003 with permission.)

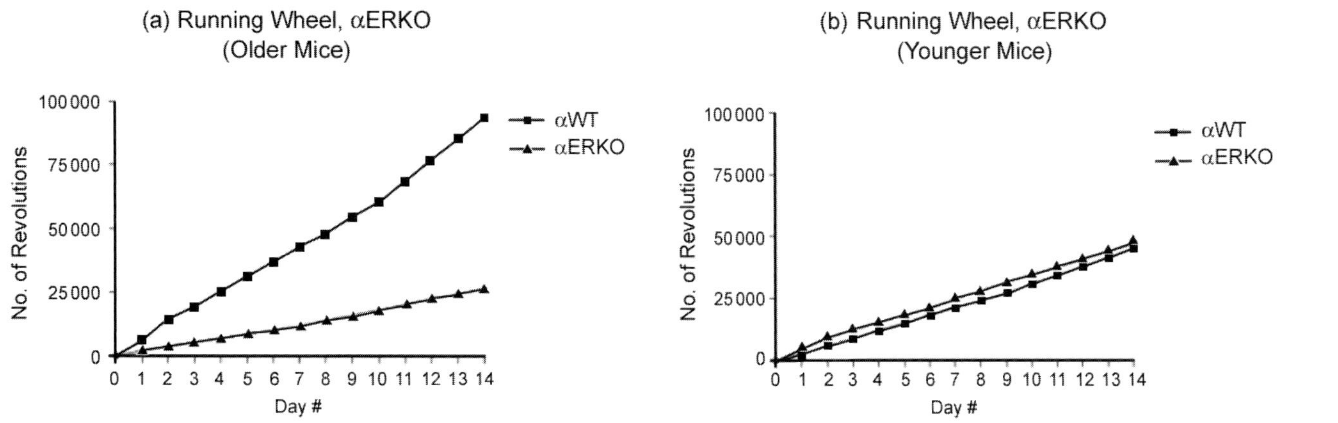

FIG. 4. Arousal, as measured by voluntary motor activity. (a) Female mice with a null mutation for αER (αERKO) showed significantly less activity in their running wheels than their wild-type littermate controls on the same genetic background (αWT). (b) Younger female mice (15 weeks of age) did not show this phenotype, for reasons we do not understand. Female mice with knockout of the βER were not different from their wild-type littermate controls. (From Garey et al 2003 with permission.)

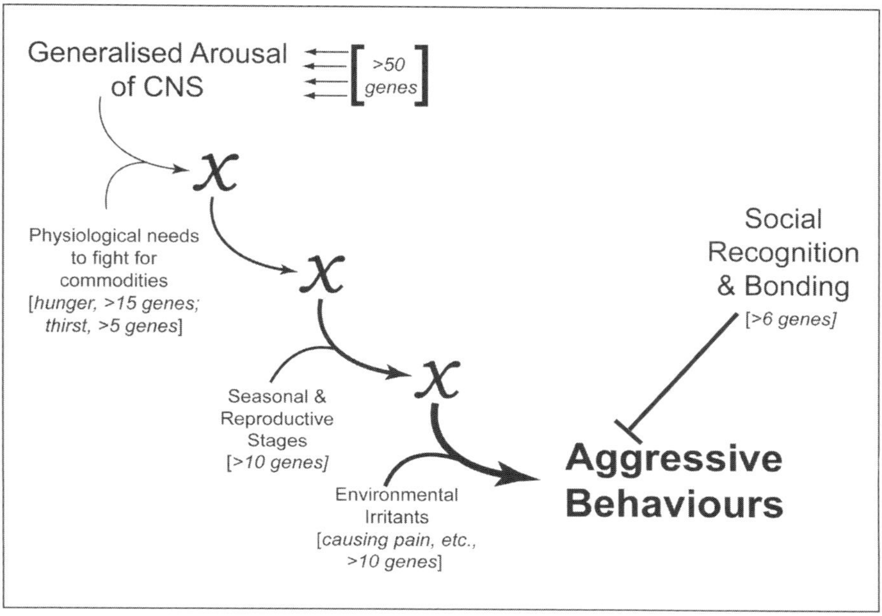

FIG. 5. A chart illustrating the diverse manner in which large numbers of genes bear on the expression of aggressive behaviours. The causal routes for these genes are multiple, both direct and indirect. They range from the effects of genes supporting (or reducing) fundamental arousal of the CNS through the genes involved in the ameliorating effects of social recognition and bonding.

differences in aggressive behaviours (See chapters in Pfaff et al 2002). In addition, functional genomic studies of aggression *qua* aggression had identified at least 36 other genes as of 2002 (Maxson & Canastar 2003). In all cases, we understand that mice with genetic alterations increasing aggressive behaviour frequency or intensity are reacting differently to specific aspects of their environments, rather than displaying aggression *in vacuo* (Nyberg et al 2004).

A much more expansive and forward-looking point also comes into view. For charting all significant gene/behaviour causal routes, we must conceive of the genes involved in every elementary subfunction fundamental to aggression, as well as the gene products centrally involved in regulating the physiological or environmental variables to which aggressive behaviours constitute possible responses. A subset of these, illustrating the theoretical roles of more than 96 genes, are sketched in Fig. 5. Thus, large numbers of genes participate in long chains of causal events leading to aggression deriving from both physiological and environmental triggers. Viewed this way it is easy to understand that, correspondingly, they will have much different causal routes. Finally, Fig. 5 gives a concrete and realistic example of how gene/environment interactions work.

References

Choleris E, Gustafsson JA, Korach KS, Muglia LJ, Pfaff DW, Ogawa S 2003a An estrogen-dependent four-gene micronet regulating social recognition: a study with oxytocin and estrogen receptor-alpha and -beta knockout mice. Proc Natl Acad Sci USA 100:6192–6197

Choleris E, Pfaff DW, Ogawa S 2003b Similar impairments of social discrimination in estrogen receptor alpha, beta and oxytocin knockout female mice: a detailed behavioural analysis. Horm Behav 44:42

Choleris E, Little SR, Mong JA, Langer R, Pfaff DW 2003c Antisense DNA against oxytocin receptor mRNA from microspheres in the medial amygdala blocked social recognition in female mice. Society For Neuroscience 2003, abstr no:839.7

Durbak L, Pfaff D, Ogawa S 2002 Role of the estrogen receptor beta gene in testosterone inducible aggression in female mice. Society For Neuroscience, 2002, abstr no:288.4

Ferguson J, Young L, Insel TR 2002 The neuroendocrine basis of social recognition. Front Neuroendocrinol 2:200–224

Garey J, Goodwillie A, Frohlich J et al 2003 Genetic contributions to generalised arousal of brain and behaviour. Proc Natl Acad Sci USA 100:11019–11022

Kavaliers M, Colwell DD, Choleris E et al 2003 Impaired discrimination of and aversion to parasitized male odors by female oxytocin knockout mice. Genes Brain Behav 2:220–230

Kavaliers M, Agmo A, Choleris E et al 2004 Oxytocin and estrogen receptor knockout mice provide discriminably different odor cues in behavioural assays. Genes Brain Behav 3:189–195

Le T, Mirasol E, Pfaff DW, Ogawa S 2002 Role of the estrogen receptor beta gene in postpartum aggression in mice. Society For Neuroscience, 2002, abstr no:288.3

Maxson SC, Canastar A 2003 Conceptual and methodological issues in the genetics of mouse agonistic behaviour. Horm Behav 44: 258–262

McCarthy MM, Schlenker EH, Pfaff DW 1993 Enduring consequences of neonatal treatment with antisense oligodeoxynucleotides to estrogen receptor messenger ribonucleic acid on sexual differentiation of rat brain. Endocrinology 133:433–439

McGinnis MY, Lumia AR, Breuer ME, Possidente B 2002 Physical provocation potentiates aggression in male rats receiving anabolic androgenic steroids. Horm Behav 41:101–110

Mong JA, Devidze N, Frail DE et al 2003a Estradiol differentially regulates lipocalin-type prostaglandin D synthase transcript levels in the rodent brain: evidence from high-density oligonucleotide arrays and in situ hybridization. Proc Natl Acad Sci USA 100:318–323

Mong JA, Devidze N, Goodwillie A, Pfaff DW 2003b Reduction of lipocalin-type prostaglandin D synthase in the preoptic area of female mice mimics estradiol effects on arousal and sex behaviour. Proc Natl Acad Sci USA 100:15206–15211

Nomura M, Durbak L, Chan J et al 2002 Genotype/age interactions on aggressive behaviour in gonadally intact estrogen receptor beta knockout (betaERKO) male mice. Horm Behav 41:288–296

Nyberg J, Sandnabba K, Schalkwyk L, Sluyter F 2004 Genetic and environmental (inter)actions in male mouse lines selected for aggressive and nonaggressive behaviour. Genes Brain Behav 3:101–109

Ogawa S, Taylor J, Lubahn DB, Korach KS, Pfaff DW 1996 Reversal of sex roles in genetic female mice by disruption of estrogen receptor gene. Neuroendocrinology 64:467–470

Ogawa S, Washburn TS, Taylor J, Lubahn DB, Korach KS, Pfaff DW 1998a Modifications of testosterone-dependent behaviours by estrogen receptor-alpha gene disruption in male mice. Endocrinology 139:5058–5069

Ogawa S, Eng V, Taylor J, Lubahn DB, Korach KS, Pfaff DW 1998b Roles of estrogen receptor-alpha gene expression in reproduction-related behaviours in female mice. Endocrinology 139:5070–5081

Ogawa S, Choleris E, Pfaff DW 2005 Genetic influences on aggressive behaviors and arousability in animals. Ann NY Acad Sci 1036:257–266

Pfaff DW 2005 Brain arousal and information theory: neural, hormonal and genetic analyses. Harvard University Press

Pfaff DW, Arnold A, Etgen AM, Fahrbach SE, Moss RL, Rubin RR (eds) 2002 Hormones, brain and behaviour. San Diego, CA, Academic Press

Simon NG 2002 Hormonal processis in the development and expression of aggressive behavior. In: Pfaff D, Arnold A, Etgen A, Fahrbach S, Rubin R (Eds) Hormones, brain and behavior. Academic Press/Elsevier, San Diego, vol I, p 339–393

DISCUSSION

Brembs: Your odour recognition experiment reminded me a lot of a dishabituation experiment. The control experiments were just standard non-social habituation and dishabituation, and they were perfectly normal.

Pfaff: That is true, and there is also nothing wrong with the experimental animals' olfactory recognition because they can find buried food.

Suomi: To what extent does the general model you presented apply to other mammals, and to primates in particular? What components might be subjected to more or less inter-species variation?

Pfaff: Probably a significant degree of inter-species variation. The simplest answer would be to say that there will be tremendous variability as regards the mechanistic model of social recognition. We coalesced the quantitative data from three different kinds of gene knockouts, and put them together with a large amount of scholarship in the field of rodent neuroanatomy. This model, therefore, will probably 'travel' best among species which use olfaction and pheromones heavily. I have not seen this tested in any species other than mouse. We have built upon this by going into the amygdala with antisense DNA against the oxytocin receptor, and we have duplicated those results. But I have a feeling that this particular genetic/functional module will be less important when we are dealing with primate species.

On the other hand, many of the neural, hormonal and genetic subjects I have dealt with in molecular endocrinology, and many of the mechanisms of primitive elementary behaviours such as sex behaviours, have considerable numbers of conserved elements. In chapter 8 of my 1999 book (Pfaff 1999), I identified 17 or 18 different mechanistic domains in which certain neuroendocrine mechanisms have been conserved between the animal and human brain. I don't think that those neurons and those molecular adaptations will be absent from the primary brain; they will simply be unimportant. With respect to the neurobiology of arousal, the mechanisms themselves are remarkably well conserved across vertebrate species. Noradrenaline, dopamine and histamine are universal. The gene for narcolepsy, discovered in dogs, is true in the mouse and explains narcolepsy in humans, for example. There are many other examples as well. The impact of these on any given behaviour, however, will vary according to the species. There is a well known law within behavioural biology, namely that there can be inverted 'U' curves showing the impact of the level of arousal on the performance of any given task. The locations of the optimum are not only going to be different across species, but they are also going to be different among individuals within a species.

Ferris: Going back to the older work of de Wied, the concept of arousal was key. When he was giving peripheral injections of vasopressin and claiming it was improving memory consolidation, many of the early data would suggest that it was enhancing arousal and these animals were able to better attend. The social memory model you have for oxytocin, ERα and the amygdala is comparable to that described for vasopressin in the septum of rats. Robert Dantzer did wonderful studies in the late 1980s giving vasopressin antagonist directly into the septum of rats and depressing social memory. A novel intruder would always attract the attention of the resident; however, if the resident had prior experience of the intruder less time was spent investigating. Rats treated with vasopressin antagonist acted like they never saw the same intruder before.

Koolhaas: Did you test these animals in a non-social learning paradigm?

Pfaff: Yes, they were not impaired.

Koolhas: The reason I am asking is that Hans-Peter Lipp tested several hundred different knockout mice in a variety of learning paradigms. He came to the depressing conclusion all these knockouts are doing is creating subliminal brain damage.

Pfaff: I've learned to distrust global statements like this. Randy Nelson will remember the days when someone said that all knockout animals are aggressive. This is clearly not true.

Koolhas: We need proper controls.

Pfaff: I think so.

Keverne: If we are talking about learning here, we have to be careful to distinguish short-term memory from more long-term effects. Did you test them the next day to see whether they had retained the memory that they had made?

Pfaff: We need to do a lot more of this. The only reason we are starting to do this now is because of a partial failure. The studies I have shown were in females, and we were trying to use an ERβ agonist in order to improve behaviour, so we had to render the behaviour suboptimal to begin with. But at first our paradigm did not work with males. With the male mice we are having to extend the inter-exposure interval in order to get their performance to be lousy, so we can then try to improve their performance.

Nelson: Have you gone forward and looked at this in a functional way? For example, do these mice show the Bruce effect? This is where a pregnant female mouse when exposed to a male mouse that is not the stud aborts the litter and mates with the new male. This phenomenon depends on social memory.

Pfaff: No, we haven't tried to extend the studies this way, although it is a good idea.

Keverne: This again is a long-term memory.

Hinde: Carrying on from Barry Keverne's point, if you do the opposite and make the interval short, you may get an increase from exposure to exposure—an excitation increase.

Pfaff: That's easy to do, because the experiments can be set up precisely. The nature of the experimental design is that we can turn over these studies rapidly. We can go from 30 s via logarithmically spaced intervals up to 24 h.

Olivier: Have you checked your animals for the level of anxiety or fear? Aggression is sometimes correlated with these phenomena. Perhaps your animals are more or less fearful.

Pfaff: The three main knockouts I have talked about are dealing with the genes coding for ERα, ERβ and oxytocin. For ERα we have done anxiety assays. We want to have more than one assay converging in order thoroughly to substantiate any given intellectual interpretation. We have used the elevated plus maze and the light–dark emergence test, and also the open field. The gene knockout results for ERα, female or male, show very small differences. There is not enough difference to explain the all or none aggressive behaviour differences.

Olivier: What is the background?

Pfaff: C57. We breed back over 10 generations to C57. The ERβ knockouts are more anxious. We hypothesize that this is because of oestrogen working on serotonergic neurons in the midbrain raphe. We haven't tested this. The reason we went after the oxytocin knockout data with much vigour was that a generation of behavioural biology and neuropharmacology had predicted a certain kind of phenotype. The initial knockout studies didn't show that much of a result on aggressive behaviour. We wanted to test under more complicated circumstances to bring out the kind of genomic functionality for which the oxytocin gene had evolved. So we devised and used a seminatural environment. There was no explicit indication of anxiety. Instead the oxytocin knockout females 'take charge'. The world-type littermate controls are like 'the second team'. Typically, an oxytocin knockout animal will decide where the communal nest is going to be. She will take all the nesting material and put it in one corner, and this works for about 2 days. Then a second oxytocin knockout female will take over from the first, and move the nest again, and she will be the alpha female for the rest of the time.

Suomi: What happens to the first female?

Pfaff: She doesn't get killed. The only animal to be wounded is the male. We start with all-female groupings because we wanted to get away from the old way of doing this kind of study, which is to set up the males and put in a female. We wanted all female groups. The gene–behaviour relations of female mice are somewhat more precise. If we put in the male unprotected, he gets attacked. If you protect the male by putting him in a cage, where the females can't get at him, and if the food is on top of the cage the only thing that happens is that the females steal his food. There is no indication of increased anxiety. When we do the elevated plus and light–dark assay, there is no indication of increased anxiety. I would refer you to the paper by Margaret McCarthy in rats (McCarthy et al 1996) where she concluded that in the presence of oestrogens, oxytocin can act as an anxiolytic. This is because oestrogens turn on the oestrogen receptor.

Keverne: You showed a slide giving a cascade or layered effect of 96 genes. Clearly, we don't expect these behaviours to be simple, but what is surprising is that just one of these genes can regulate many of those tiers, acting like a synchronizer.

Pfaff: What you have pointed out with such perspicacity is the opportunity for a calculus, in which an individual gene can have quantitative modulatory effects at a number of different levels. This is related to another point: in terms of discovering individual gene–behaviour relationships we are going to have a lot of fun: I had left some Greek Xs in the algorithm, because as we go from the upper left part of the model to the lower right we were dealing with a mathematician would call an 'increasing function'. This is a sloppy term — what we don't know is whether some of the interactions are additive, or multiplicative, or work by power law. What are the quantitative translations? Your point, that an individual gene is going to play at several levels, is valid. The question then is how these levels interact with each other quantitatively, and this will make for 20 years' funded work!

Suomi: This also suggests the danger of jumping to premature conclusions: i.e. that you have the whole story when you demonstrate effects of a single gene at all these different levels.

Koolhaas: There is even the question of whether these are the levels, or can we think of other levels with different genes? How is the system or cascade organized?

Pfaff: I wanted to put only enough levels in the model to illustrate the potential complexity of the system at a glance.

R Blanchard: I take it you have run the animal's tendency to switch to novel stimuli other than olfactory stimuli? There is a tendency for animals to select novelty over familiarity in other situations.

Pfaff: Martin Kavaliers has left the lab, and I think he is doing this now at the University of Western Ontario.

R Blanchard: I am always curious about the notion of arousal and its specificity. In a certain sense, we have different kinds of arousal, ranging from the very general basal wake–sleep arousal to a situation where an animal identifies something as potentially dangerous. These reflect some differences in the specificity and also the level of arousal. How do you conceptualize arousal in general versus specific arousal?

Pfaff: In the form of an equation. There was a false dichotomy in arousal neurobiology that lasted for about 40 years. On one side were neurophysiologists who worked from about 1940–1975. They tended to think of arousal pathways as a single monolithic force which would emanate from the pons and the midbrain and flood the cerebral cortex with electrophysiological activity which would lead to arousal of brain and behaviour. Set up against this point of view was a whole generation of precise behavioural work in which they showed that there were so many different facets of arousal in varied behavioural situations, and in different species, that the concept of arousal might be Balkanised. A couple of them said that it was so subdivided that we shouldn't use the word arousal any more.

I felt that these polar opposites were fighting with each other needlessly. If you think of this in the form of an equation, the arousal of any vertebrate at any moment is a compound increasing function (mathematically a slippery term) of a generalized arousal force acting together with a large number of specific arousal forces which depend on differently biologically regulated motivations, including pain and fear. In this formulation we have not yet guaranteed the exact nature of the metric by which different specific arousal forces (such as fear) impact the overall equation. The equation gives us the general form of the solution for the problem, and resolves the old dichotomy. This gives us a precise way of thinking about it, and now the question is how to turn this into an experimental reality. What we are doing now is asking how two specific arousal forces act on each other quantitatively in order to increase generalized arousal by our high-throughput assay, and by measuring specific forms of arousal, namely sex arousal and fear.

Hinde: The word 'arousal' becomes meaningless when you use it in this loose way. There would be something to be said for confining it to what I believe to be the original sense — namely 'general' arousal. If that is shown to be inadequate, then the term could be dropped.

Pfaff: Yes, it has too much of a history. I wish I could think of a different term.

Nelson: We don't need to reinvent the word 'arousal'. We just have to do as you have done, which is to carefully define it operationally. It is a long-standing hypothetical construct, similar to motivation, perception and attention. We can't measure any of these things directly. We can only measure performance or output in some defined way.

Pfaff: I appreciate that remark. A sweet way of thinking about it is that the term 'information' was a slippery, much-maligned term. Everyone knew that it was important, but no one knew what it was until 1948, when a paper by Shannon & Weaver gave an intuitive definition of information coupled to an equation which matched the definition, and the equation turned out to be useful. Now we have Shannon information, which has been applied to biology extensively, and we are still using Shannon information calculations 56 years hence. Shannon information was really a translation into electronic circuit theory of the concept of entropy from thermodynamics. We would be happy (and lucky) if we could give a precise operational definition of arousal in a simple animal, and also find a precise measure of it that turned out to be useful, such that we could then turn this term from a sloppy term into a precise one. When I look in the dictionary I haven't found a substitute.

C Blanchard: In the activity situation in the home cage, were the mice male or female?

Pfaff: Female. We wanted to link that experiment to what we already knew about gene–behaviour relations in female mice.

C Blanchard: Were the female tests of aggression done in young or old animals?

Pfaff: In older animals. I can tell you that both the functional genomics of aggression and of arousal depend on the age of the animal. Also, the neuropharmacology of generalised arousal is not the same in males and females. If we give a histamine 1 receptor blocker to males and females, the results in the two genders are different (Easton et al 2004).

Brodkin: I have a question that is slightly off topic. In this cascade of behaviours that underlie other behaviours, I was reminded of a topic that came up earlier — the relationship between fear and aggression. I know that, generally, if you increase fear in a rodent, you can decrease that rodent's aggression. If you expose a mouse to a cat, it becomes less aggressive, for example. From the standpoint of these types of behavioural experiments, increasing fear decreases aggression. But if you use genetics to address this question, for example if you knock out a gene that decreases fear, do you always increase aggression, and vice versa? Is there a clear relationship? Are fear and aggression always negatively correlated.

Lesch: I will address this topic in my paper (Lesch 2005, this volume). The answer is clearly no. There are certain genes that can be knocked out which increase fear or anxiety-like behaviour and decrease aggression at the same time, and other genes which increase both.

Brodkin: It seems that, at least genetically, there is a complicated relationship between fear and aggression. Intuitively, we could imagine that increasing fear might decrease aggression, but, alternatively, it might make sense that, in some situations, increasing fear might cause a fight-or-flight reaction that might lead to fighting.

Suomi: You don't have to do a knockout to demonstrate the phenomenon behaviourally. There are a variety of other manipulations that will yield essentially the same thing.

Keverne: With regard to the different levels at which a given gene can have an effect, on one of your slides I got the impression that the stimulus value of the animals changed. Diane Robins, do these oestrogen receptors (ERs) have any effect on either the liver or the salivary glands, for changing the lipocalin proteins?

Robins: I don't know about those specifically, but since this is a 'whole body' ER knockout one would expect them to be affected if they are ER-regulated.

Keverne: The MUPs, which are lipocalins, and the lipocalins in the salivary gland are all important for communicating an animal's individuality to itself — it needs to know 'I am me' so it can tell something which is different.

Pfaff: I would be the last person in the world to argue against your point! In 1973 we published the effects of oestrogen on tactile sensitivity (Kow & Pfaff 1973). Therefore an animal could have differential tactile sensitivity, including sensitivity of C fibres, which would modulate pain, according to the hormonal state. Further, my earliest neurophysiology was actually in the olfactory system where I was trying to discover the effects of pheromones. David Moulton was an olfactory physiologist at Clark University in Worcester (MA), and he made the nice point in a review article that when hormones affect responses to odours, it may not be because of their effects on neurophysiology; it may be because of effects on the volume and viscosity of the nasal mucosa (Moulton 1976). This yields an example in which the time interval for the passage of a pheromone across the sensory surface could be influenced by hormonal status.

Keverne: Of course. There are oxytocin receptors in spinal cord which are also influenced by oestrogen and affect somatosensory stimui.

References

Easton A, Norton J, Goodwillie A, Pfaff DW 2004 Sex differences in mouse behavior following pyrilamine treatment: role of histamine 1 receptors in arousal. Pharmacol Biochem Behav 79:563–572

Kow LM, Pfaff DW 1973 Effects of estrogen treatment on the size of receptive field and response threshold of pudendal nerve in the female rat. Neuroendocrinology 13:299–313

Lesch K-P 2005 Serotonergic gene activation in mice: models for anxiety and aggression? In: Molecular mechanisms influencing aggressive behaviours. Wiley, Chichester (Novartis Found Symp 268) p 111–146

McCarthy MM, McDonald CH, Brooks PJ, Goldman D 1996 An anxiolytic action of oxytocin is enhanced by estrogen in the mouse. Physiol Behav 60:1209–1215

Moulton DG 1976 Spatial patterning of response to odors in the peripheral olfactory system. Physiol Rev 56:578–593

Pfaff DW 1999 Drive: neurobiological and molecular mechanisms of sexual motivation. MIT Press, Cambridge, MA

General discussion I

Ferris: As we try to translate from an animal model to the human condition, choosing an ethologically valid model is critical. Much of the early work in aggression was spent on experimentally manipulating animals to get an aggressive response. If you injured or electrically shocked an animal you would get biting attacks (usually directed to the handler). Consequently, drugs were tested that could stop this kind of aggressive behaviour. As you would expect, antipsychotics and tranquilizers proved to be effective but none specific. Essentially the animal was being immobilized by the drug treatment and unable to attend to its environment. The better approach is to come up with an ethologically valid model to help us understand the normal aggressive behaviour of animals and the natural stimuli that elicit biting attacks. When you understand the normal neurobiology responsible for agonistic behaviour then you can look for environmental conditions that might alter the brain and behaviour. That said, we then turn to the human aggression picture, adopting the notion that it is a pathology of impulsivity, violence and cognitive dysfunction secondary to some other primary mental problem. How do we model this? We are looking at ethologically valid conditions in hamsters, rats or monkeys, but does this really translate to understanding aggression in humans? Should we be trying to come up with a model that may occur naturally to push an animal across the spectrum of normal behaviour to inappropriate excessive aggression? For example, in dominance hierarchies perhaps it is best to study the subordinate animals and how their behaviour and neurobiology change during social subjugation. The underlying neurobiological principles governing offensive aggression may be similar across species but altered after traumatic psychosocial and environmental experiences. Understanding how the brain in laboratory animals adapts to negative events may help us to better understand the psychopathology around human impulsivity and violence.

Suomi: For just about any aspect of development, one can describe species-normative patterns that generate a basic developmental trajectory and then describe individual differences in terms of deviations from that general pattern. This allows you to select subsets of individuals that might conform to certain aspects of the human situation, and it allows one to compare these with other developmental trajectories at any level of analysis one wishes to examine. Presumably one could do the same with aggression.

Ferris: You do this with Rhesus monkeys, and have colonies of hundreds of animals. This work on non-human primates brings us closer to the human condition and greatly improves translation. However, what can scientists that study mice, rats and hamsters bring to the table?

Koolhaas: We use the wild-type laboratory rat, which has a large variation in aggressive behaviour. With a proportion of these animals, if you give them repeated winning experiences, they strongly reduce all their introductory behaviour and in my view they develop a violent kind of behaviour, to the point that they even attack anaesthetized opponents. They don't take the pinching bites; they make tears of several centimetres. They become killers. I think here we have a model to address pathological forms of aggressive behaviour.

Keverne: It sounds to me as if there is an element of learning coming in there. When I say 'learning', I mean it in a biological sense. It can be by second-order association with reward events.

Koolhaas: We can demonstrate in a variety of situations that with repeated experience the animals don't pay attention any longer to environmental cues. This is not only social cues, but all kinds. They tend to anticipate the situation.

Martinez: Is this in 100% of cases?

Koolhaas: The top 30% of the population does this.

Martinez: What were the ones who became killers like before?

Koolhaas: They were perfectly normal in all aspects.

R Blanchard: We saw a similar sort of thing in our social groups. In our visible burrow systems we have seen one or two residents that were very dangerous to male intruders and also male members of their own groups: their attack was not constrained, and abdominal wounding and even death resulted. This was apparently a form of psychopathology in these rats, in that it was maladaptive. Under normal conditions they would have been likely to kill their own male offspring as well.

This brings up an issue. Part of our problem is that we want to understand human aggressive psychopathology. We need to have a better understanding of these kinds of behaviours in terms of human ethology. If we are to use behavioural descriptions in animals and relate these to people, we also need to have behavioural descriptions and analyses of anger and aggression much more clearly laid out in humans. Unfortunately, this is really hard to do, so we tend to do it indirectly through batteries of tests. It is a big problem.

C Blanchard: These particular 'pathological' rats apparently got that way without any kind of unusual experience. They had not been treated any differently than the others, and had lived much of their adult lives in single housing, yet they freely bit vulnerable areas of their opponents, something that you cannot easily get a normal rat to do. We disposed of these guys immediately, but in retrospect we should have kept them and bred them. But the point of the story, in relating animal to human aggression, is that this is a two category by two category comparison: animals and people; normal and abnormal. We have to study all four categories. At the moment we tend to study normal aggression in animals and pathological aggression in people. This is likely to be suboptimally productive.

Keverne: You said these rats had no experience. Were they brought up in social isolation? This is a strong promoter of this kind of aggression.

Martinez: Rats don't become aggressive by being isolated; this is just in mice.

Brodkin: We were talking about types of aggression. There is another distinction in humans between impulsive aggression and more premeditated, planned aggression. Are most animal models really models of impulsive aggression? Is there a good animal model of premeditated aggression? Would that be similar to predation?

Brembs: Chimpanzee warfare.

Suomi: In the normative range of behaviour in field populations there are plenty of examples of both types. It is possible to subdivide and come up with precise operational definitions of different forms that are shown by different individuals under varied circumstances, with different underlying biological correlates. For example, dominance-related aggression seems to be related to testosterone, whereas impulsive aggression seems to involve serotonin more.

C Blanchard: People who show impulsive aggression tend also to display other impulsive behaviours.

Manuck: I think Dr Blanchard's point is true, especially where impulsive aggression has been found associated with variation in brain serotonergic activity (as Dr Suomi just suggested). For instance, dysregulation of central serotonergic function in nonhuman primates occasions an uninhibited approach to social strangers (Fairbanks et al 2001, Manuck et al 2003) and, in contexts devoid of social interactions, impulsive risk-taking (e.g. spontaneous leaping at dangerous heights among forest trees) (Higley et al 1996, Mehlman et al 1994).

In a non-patient community sample, we have found that men who report having engaged in high levels of aggressive conduct over the course of their lives show reduced central serotonergic responsivity (a blunted prolactin response to the serotonin agonist, fenfluramine) when compared to their more placid counterparts. Even in the absence of overt psychopathology, difficulties of adjustment may be seen in the life histories and associated personality characteristics of these 'normatively' aggressive men, including a greater prevalence of prior substance-use disorders, incomes incommensurate with earlier educational attainments, marital instability (fourfold higher divorce rates), and higher scores on standardized measures of hostility and impulsive disposition (Manuck et al 2002).

My second point has to do with the suitability of different animals as models of human aggressiveness. I do not know much about rodent models (which seem quite popular in aggression studies), but in other areas of research with which I am familiar — particularly the behavioural exacerbation of coronary artery disease — the rat is not a preferred model due to dissimilarities (from humans) in both the nature of its sociality and the pathobiology of arterial lesions. I wonder if some of the same considerations bear on the utility of different animal models of aggression — at least for the study of individual differences — with some animals corresponding more closely to humans in regard to their regulatory neurobiology and dimensional structure of predisposing behavioural traits.

Brembs: As I am a biologist and researcher in invertebrate neurobiology, it will probably come as no surprise to you that I think that a multitude of models will serve best. From an evolutionary point of view, you will find commonalities and differences that are not only interesting in themselves, but which will also lead you down paths and make you think about things in human aggression that you would never have thought about. There is value in using several different model systems. For me as a biologist, I don't know the statistics about how relevant what we call human pathologies of aggression are in terms of societal burden (damage, injuries and deaths). What we could discuss is whether wild-type aggression is accounting for the vast extent of what we are trying to get rid of. A third point is that I was surprised when we were talking about human aggression that the term provocation didn't arise. In my own experience I have known people whose favourite pastime was to pick fights. They would never be the first one to punch, but they had a weird way of finding susceptible people and provoked them until a fight would break loose. An intruder in an animal situation could be the aggressor, even though he will be attacked. Perhaps we should be considering provocation as well as physical aggression itself.

References

Fairbanks LA, Melaga WP, Jorgensen MJ, Kaplan JR, McGuire MT 2001 Social impulsivity inversely associated with CSF 5-HIAA and fluoxetine exposure in vervet monkeys. Neuropsychopharmacology 24:370–378

Higley JD, Mehlman PT, Poland RE et al 1996 CSF testosterone and 5-HIAA correlate with different types of aggressive behaviors. Biol Psychiatry 40:1067–1082

Manuck SB, Flory JD, Muldoon MF, Ferrell RE 2002 Central nervous system serotonergic responsivity and aggressive disposition in men. Physiol Behav 77:705–709

Manuck SB, Kaplan JR, Rymeski BA, Fairbanks LA, Wilson ME 2003Approach to a social stranger is associated with low central nervous system serotonergic responsivity in female cynomolgus monkeys (*Macaca fascicularis*). Am J Primatol 61:187–194

Mehlman PT, Higley JD, Fauener I et al 1994 Low CSF 5-HIAA concentrations and severe aggression and impaired impulse control in nonhuman primates. Am J Psychiatry 151:1485–1491

Molecular architecture of pheromone sensing in mammals

Catherine Dulac

Department of Molecular and Cellular Biology, Howard Hughes Medical Institute, Harvard University, 16 Divinity Avenue, Cambridge, MA 02138, USA

Abstract. Pheromones have evolved as a discrete class of secreted chemicals that signal the sex and social status of an individual and that promote coordinated motor programs and endocrine changes essential for breeding and aggression. The highly reproducible and species-specific character of the response to pheromones offers a unique experimental system to elucidate the coding of sexual and social information within the brain, and to provide new insights into the molecular and cellular basis of information processing leading to specific behaviours.

2005 Molecular mechanisms influencing aggressive behaviours. Wiley, Chichester (Novartis Foundation Symposium 268) p 100–110

The nasal cavity of most mammals contains two functionally and anatomically independent sets of olfactory neurons. Olfactory neurons of the main olfactory epithelium (MOE) are located in the posterior recess of the nasal cavity and detect the large variety of small volatile chemicals that are carried by the air during breathing. The sensory information generated in the MOE is transmitted to the main olfactory bulb (MOB) and processed by multiple cortical and neocortical centres of the brain, leading to the cognitive and emotional responses to a smell (Fig. 1, right). Olfactory stimuli elicit adaptive behavioural responses that are largely shaped by sensory experience. In contrast, pheromones are primarily detected by neurons of the vomeronasal organ (VNO), a bilateral and tubular-shaped chemosensory structure of the ventral nasal septum. The VNO lumen opens into the nasal groove of the nasal cavity where it has access to aqueous soluble pheromone signals carried by the nasal mucus. VNO neurons send fibres to the accessory olfactory bulb (AOB) which projects to the medio-cortical amygdala and in turn to discrete loci of the hypothalamus involved in reproductive and aggressive responses (Fig. 1, left). The role of the VNO in the pheromone-evoked response of rodents has been suggested by surgical ablation of the VNO neuroepithelium and its projection to the AOB that resulted in impairment in mating and aggressive behaviours, and in pheromone-induced endocrine responses.

Molecular biology of pheromone and odorant detection

How is the pheromone information detected by chemosensory receptors and translated into appropriate behavioural and hormonal responses? How is the diversity and the specificity of the pheromone response accomplished? The initial event of chemosensory detection requires the activation of specific receptors by chemosensory signals and the translation of receptor activation into changes in membrane potential. Remarkably, molecular studies have found that MOE and VNO sensory neurons utilize evolutionarily independent signal transduction mechanisms to mediate chemosensory signalling (Fig. 1). In the MOE, specific recognition of odorants is achieved by the expression of a large family of about a thousand G protein-coupled odorant receptors (ORs) and neuronal activation is mediated by a cyclic nucleotide-dependent signal transduction cascade (Buck & Axel 1991). In contrast, VNO signalling involves several receptor families molecularly unrelated from that of the OR. Differential screening of cDNA libraries constructed from individual neurons from rodent VNO has led to the isolation of two large and independent families of vomeronasal receptor genes (VR), the V1Rs and V2Rs, that encode VNO-specific G protein-coupled receptors (GPCRs) (reviewed in Dulac & Torello 2003). V1Rs and V2Rs are unrelated both to each other and to the olfactory receptors (ORs) and are thought to represent two distinct families of putative pheromone receptors. Each family comprises several hundred putative pheromone receptor genes that are expressed by two spatially segregated populations of VNO sensory neurons (Fig. 1). Neurons lining the luminal (or apical) half of the VNO neuroepithelium coexpress V1Rs as well as the G protein alpha subunit $G_{\alpha i2}$. In contrast, the basal half of the VNO epithelium contains neurons that are V2R- and $G_{\alpha o}$-positive. Within each area of the VNO, the expression of specific V1R or V2R receptor genes is confined to small (0.1–1%) and non-overlapping subpopulations of VNO neurons, suggesting that individual VNO neurons are likely to express only one receptor. This result is consistent with a model of vomeronasal signalling according to which the binding of pheromones to specific receptors leads to the activation of distinct subpopulations of VNO neurons, thus permitting neuronal discrimination of pheromonal cues. VNO sensory fibres reaching the AOB remain segregated according to their origin from the apical or basal sides of the VNO, with $G_{\alpha o}$-positive fibres projecting to the posterior half of the AOB and $G_{\alpha i2}$-positive fibres reaching the anterior portion of the AOB. The expression of molecularly divergent pheromone receptors by these two spatially segregated populations of sensory neurons may indicate that the vomeronasal system in fact comprises two separate functional units in order to detect and process pheromone signals.

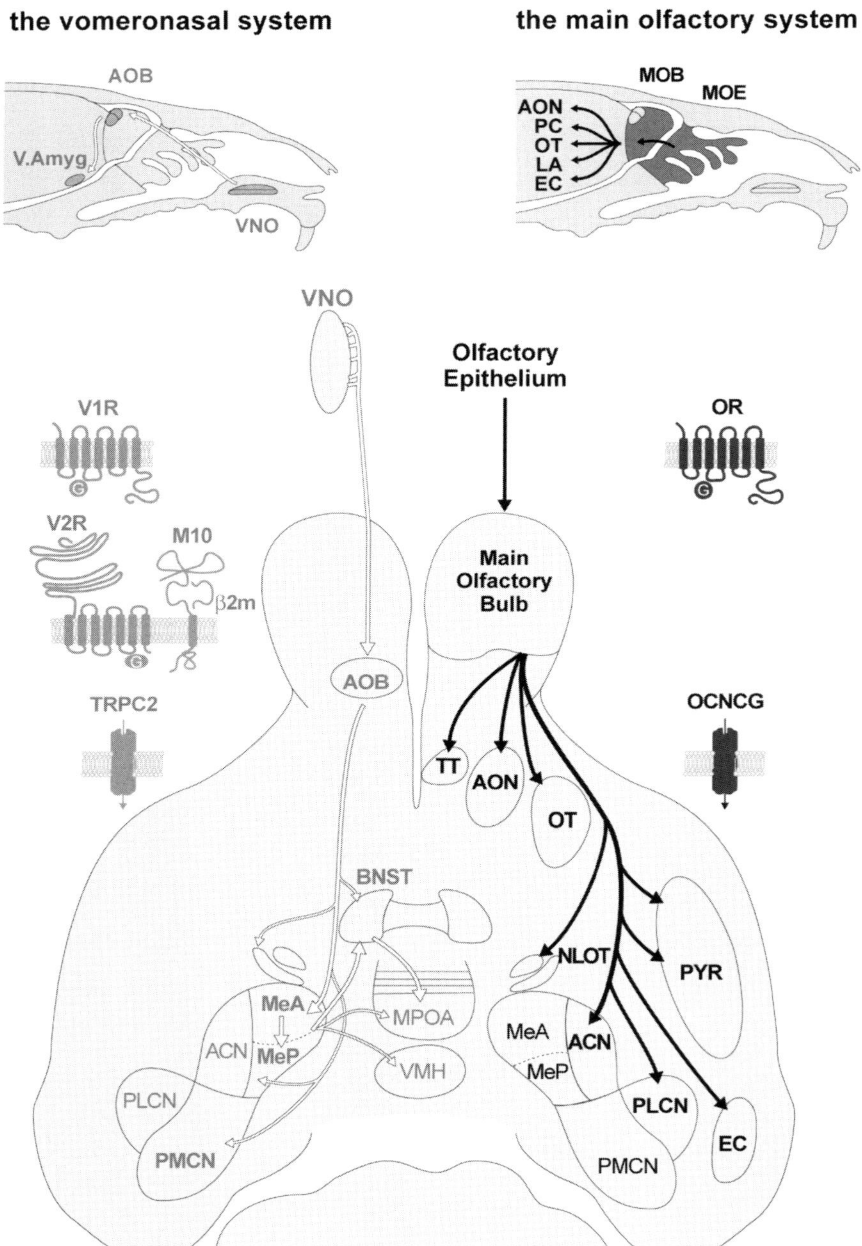

the vomeronasal system
the main olfactory system
AOB
V.Amyg
VNO
MOB
MOE
AON
PC
OT
LA
EC
VNO
Olfactory Epithelium
V1R
OR
V2R
M10
β2m
Main Olfactory Bulb
AOB
TRPC2
OCNCG
TT
AON
OT
BNST
NLOT
PYR
MeA
MPOA
ACN
MeP
VMH
MeA
ACN
MeP
PLCN
PMCN
PLCN
EC
PMCN

Coexpression of MHC-like molecules with V2Rs in the VNO

Recently, we and others have uncovered the expression of two families of VNO specific non-classical major histocompatibility complex (MHC) molecules, M1 and M10 (Ishii et al 2003, Loconto et al 2003). Individual M10 genes are expressed in subpopulations of VRNs in the basal region of the VNO neuroepithelium, such that individual neurons express only one or a few M10s. Intriguingly, expression of a given M10 is positively correlated with the expression of specific V2Rs; thus, the selective association of specific M10:V2R combinations might define populations of molecularly distinct VRNs.

What is the functional role of M10 in the VNO? β2-microglobulin (β2m) a molecule which stabilizes the three dimensional structure of many MHC molecules at the cell surface, is co-expressed with M10 in VNO neurons (Loconto et al 2003). Moreover we were able to demonstrate that β2m, M10 and V2R form a multimolecular complex that is localized to the dendritic tip of the receptor neurons (Loconto et al 2003). The fact that M10 is not expressed in embryonic or early postnatal stages, and that M10 and V2R are co-localized to the dendritic tips of VNO neurons (the site of pheromone detection), suggests that M10 is directly involved in the process of pheromone detection.

Previous experiments have shown that V2Rs expressed in heterologous cell types are inefficiently expressed at the cell surface. To investigate the functional role of the M10:V2R interaction, β2m and an M10:V2R pair were co-expressed in a mouse spermatogonia cell line. Surprisingly, both the M10 and the V2R were efficiently targeted to the cell membrane, suggesting a role for M10 in V2R trafficking. In support of this hypothesis, immunohistochemical analysis of VNOs from $\beta 2m^{-/-}$ mutant mice shows that localization of V2Rs to the dendritic tip of the VNO neurons is severely compromised (Loconto et al 2003).

Thus, both *in vitro* and *in vivo* experiments demonstrate that M10 is involved in V2R transport, suggesting that M10 is exclusively involved in receptor trafficking. Accessory molecules required for chemosensory receptor transport have been

---

FIG. 1. Functional, anatomical, and molecular segregation of the two mammalian olfactory systems. Olfactory sensory neurons in the main olfactory epithelium (MOE) are specialized in detecting small volatile odorants. The olfactory information is transmitted from the MOE to the main olfactory bulb (MOB), then to distinct brain nuclei that form the primary olfactory cortex and that include the anterior olfactory nucleus (AON), the piriform cortex (PC), the olfactory tubercle (OT), the entorhinal cortex (EC) and the lateral amygdala (LA). In contrast, the information provided by pheromone signals is processed by a distinct neural circuit. Pheromones are detected by sensory neurons in the vomeronasal organ (VNO), a bilateral tubular structure in the anterior region of the nasal cavity. VNO axons project to the accessory olfactory bulb (AOB), which in turn transmits sensory information to the vomeronasal amygdala (VA) and then to specific nuclei of the hypothalamus (H), which are involved in regulating genetically preprogrammed physiological and reproductive responses.

identified in other organisms: for example, two gene products are required for proper localization of the ODR-10 odorant receptor in *Caenorhabditis elegans*: the *unc-101* gene product encodes an Ap-1 m1 clathrin adaptor required for sorting and transport of ODR-10 to the olfactory cilia (Dwyer et al 2001), and *odr-4* encodes a novel membrane-associated protein required for proper folding, transport or localiztion of the ODR-10 odorant receptor (Dwyer et al 1998).

Alternatively, M10 might play multiple roles in V2R processing, transport and regulation of ligand specificity, similar to how the RAMPs (receptor activity modifying proteins) regulate the calcitonin-receptor-like receptor (CRLR): RAMPs have been identified as molecules that regulate the glycosylation of, are required for surface expression of, and alter the ligand specificity of the CRLR (McLatchie et al 1998). RAMP1 facilitates terminal glycosylation and membrane localization of the CRLR, and CRLR functions as a calcitonin gene-related peptide (CGRP) receptor when co-expressed with RAMP1. On the other hand, RAMPs 2 and 3 do not facilitate glycosylation of the CRLR, and when co-expressed with either RAMP2 or 3, CRLR is presented at the cell surface as adrenomedullin receptors (McLatchie et al 1998).

From signal transduction to behaviour

Unlike vertebrate photoreceptors and MOE olfactory neurons, the membrane conductance of rodent VNO neurons is insensitive to cyclic nucleotides indicating that the cyclic-nucleotide-gated channel is not the primary conductance induced by pheromone receptor activation. Instead, a model of mammalian pheromone-evoked transduction cascade has been proposed that involves a member of the TRP family of ion channels, and that would mirror the signalling pathway identified in *Drosophila* photoreceptors. Members of the TRP family of ion channels are involved in G protein-regulated and cyclic-nucleotide-independent transduction pathways in a variety of other sensory systems such as mechanosensation and olfaction in *C. elegans* and phototransduction in the fruitfly. In the *Drosophila* eye, photo-isomerization of rhodopsin activates a G_α protein of the $G_{\alpha q}$ class, which in turn triggers a phosphatidyl inositol signalling cascade, and leads to the opening of the cation-selective channels dTRP and dTRPL. Similarly, a specific TRP channel, rTRPC2 has been identified in rodents, which is exclusively expressed in VNO neurons (Liman et al 1999). The rTRPC2 protein appears highly localized to VNO sensory microvilli, the proposed site of pheromone sensory transduction. These results have suggested a direct role of rTRPC2 in the pheromone evoked-response and would be consistent with the increase in inositol-1,4,5-trisphosphate (IP3) reported in snake and rodent VNO neurons in response to pheromones. The rTRPC2 channel might represent the primary conductance activated by the pheromone signal, or it could mediate a secondary amplification or modification of the sensory response.

Genetic ablation of TRP2 in the mouse provided a new experimental system to assess the requirement of TRP2 function in VNO signalling and to directly investigate the repertoire of VNO-mediated sensory responses and behaviours (Stowers et al 2002, Leypold et al 2002). First, it appears that the TRP2 deficiency dramatically impairs the sensory activation of VNO neurons by urine pheromones, thus confirming the critical role of TRP2 in the VNO signal transduction cascade. In addition, the absence of pheromone detection mediated by VNO signalling has striking behavioural consequences. TRP2$^{-/-}$ male mice appear unable to recognize the sexual identity of their conspecifics: they fail to display the pheromone-evoked aggression toward male intruders that is normally seen in wild-type males and, remarkably, they display courtship and mounting behaviour indiscriminately toward both males and females. These data contradict the established notion that VNO activity is required for the initiation of male–female mating behaviour in the mouse and suggest instead a critical role in ensuring sex discrimination (Fig. 2). Defects in maternal aggression, male territory marking and recognition of social dominance also appear in the TRP2$^{-/-}$ mouse line.

Concluding remarks

The identification of key players of the pheromone detection apparatus, including the V1R and the V2R receptor families, the M10 and M1 receptor escorts and the TRP2 ion channel, has provided the molecular framework for our understanding of the pheromone-evoked response in the VNO. In addition, new imaging and

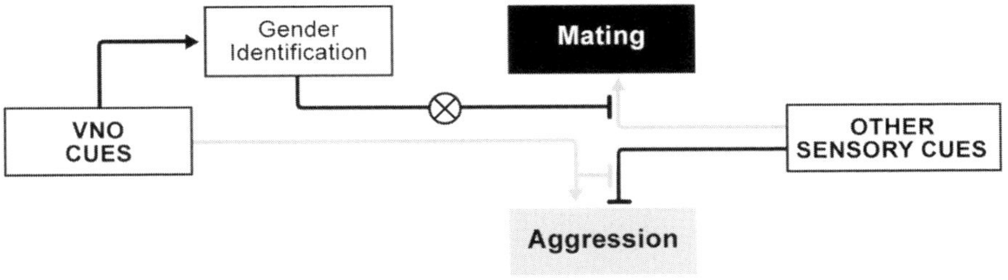

FIG. 2. Role of the mouse VNO in gender discrimination. In wild-type mice, display of gender-appropriate responses is ensured by VNO activity. The detection of male pheromones by a male results in inhibition of mating (black) and initiation of aggressive behaviours (grey). Behavioural analysis of TRPC2$^{-/-}$ males indicates that non-VNO-related sensory inputs such as olfactory, auditory, tactile, or visual cues trigger mating behaviour irrespective of the gender of the encountered mouse. So, in the absence of VNO activity, mating is the default behaviour of the male with a conspecific and aggressive responses are suppressed.

recording technologies have offered a new opportunity to examine the pheromone-evoked response of large populations of neurons in the VNO, and more recently in the brain. Both molecular and electrophysiological approaches have uncovered an unexpected diversity and complexity underlying the function of the mammalian vomeronasal system. The repertoire of pheromone receptors, for example, appears surprisingly large by comparison with the small list of behavioural and endocrine responses evidenced so far. The detection of natural stimuli lead to the activation of large populations of sensory neurons, yet, in the VNO, each receptor type displays a narrow tuning for simple compounds present at minute concentrations, and mitral cells of the AOB appear similarly tuned to pheromonal stimuli from specific sources. Finally, genetic analysis of mutant mouse lines deficient in key VNO signalling components have provided significant insights into the physiological role of the vomeronasal system in the organism, and in particular, its exclusive role in gender identification. Future studies will clearly require the ability to associate the different elements of the current puzzle: what is the significance of the myriad of chemicals detected by the VNO, how do they carry the information about sex, species, individual identity of the animal, how do they relate to the different receptor families and different mitral cell activation? Finally, the pheromones appear to act not as simple releasers of mating or aggressive behaviours, but rather as essential regulators of inputs from other sensory organs, ensuring the sex specificity of the behavioural response. Thus, in addition to the intriguing nature of a brain circuit designed to recognize sex, one wonders how the vomeronasal system impinges upon the function of the other sensory networks.

References

Buck L, Axel R 1991 A novel multigene family may encode odorant receptors: a molecular basis for odor recognition. Cell 65:175–187

Dulac C, Torello AT 2003 Molecular detection of pheromone signals in mammals: from genes to behaviour. Nat Rev Neurosci 4:551–562

Dwyer ND, Adler CE, Crump JG, L'Etoile ND, Bargmann CI 2001 Polarized dendritic transport and the AP-1 mu1 clathrin adaptor UNC-101 localize odorant receptors to olfactory cilia. Neuron 31:277–287

Ishii T, Hirota J, Mombaerts P 2003 Combinatorial coexpression of neural and immune multigene families in mouse vomeronasal sensory neurons. Curr Biol 13:394–400

Leypold BG, Yu CR, Leinders-Zufall T, Kim MM, Zufall F, Axel R 2002 Altered sexual and social behaviors in trp2 mutant mice. Proc Natl Acad Sci USA 99:6376–6381

Liman ER, Corey DP, Dulac C 1999 TRP2: a candidate transduction channel for mammalian pheromone sensory signaling. Proc Natl Acad Sci USA 96:5791–5796

Loconto J, Papes F, Chang E et al 2003 Functional expression of murine V2R pheromone receptors involves selective association with the M10 and M1 families of MHC class Ib molecules. Cell 112:607–618

McLatchie LM, Fraser NJ, Main MJ et al 1998 RAMPs regulate the transport and ligand specificity of the calcitonin-receptor-like receptor. Nature 393:333–339
Stowers L, Holy TE, Meister M, Dulac C, Koentges G 2002 Loss of sex discrimination and male-male aggression in mice deficient for TRP2. Science 295:1493–1500

DISCUSSION

Keverne: The TRP2C knockout mouse has been lacking the gene throughout embryology and postnatal development, into the adult animal. When you are talking about the gender recognition component, is it really a recognition as we understand recognition (I can recognize a male from a female by vision) or is it a recognition that is solely concerned with engaging the behaviour? The reason I ask this question is that in the 1970s I did work where I was doing differential lesions of the VNO and the main olfactory system. When the VNO is lesioned in the adult, males can still distinguish male urine from female urine.

Dulac: I am really glad you asked this. One of the most fascinating questions about gender recognition is when and how an animal starts to tell apart males and females. You could imagine that the animal has some sort of internal template, a genetically pre-programmed template, providing the intrinsic information about what is male and what is female. So if the input from the VNO does not come in, the animal is unable to discriminate. Your experiment might suggest that this is not the case. You could also imagine that at birth, all animals are in contact with their mothers, and therefore, they will always know what 'mum' is, which is a female, and they know this from very early VNO function. This would set up all the rest of the behaviour. If you remove the VNO later on this won't have much effect.

R Blanchard: Have you examined the scent marking of your knockout animals?

Dulac: We haven't, but Richard Axel has. There seems to also be a defect in recognition of scent marking. They can't recognize the dominant male and so they always act as subordinate, because they get beaten up.

Pfaff: A neuroanatomist named Mimi Helpern, at SUNY Downstate Medical School, distinguished different classes of VNO receptor cells on the basis of G protein expression. Does this distinction map onto your results?

Dulac: Yes, the V1 cells express $G_i2\alpha$ and the V2 a G_o protein. They are segregated. This is what told us that we didn't have the entire repertoire of the receptor when we just had the V1Rs, because the neurons expressing G_o could not be accounted for in terms of receptor expression.

Pfaff: You had no changes in mating behaviour, but you probably tested these animals in severely reduced environments. I would recommend that you give your animals to people who use natural or semi-natural environments. You might find out that the chain of courtship behaviours are so disrupted that mating behaviour *de facto* doesn't occur.

Dulac: You are right. It is a concern that these experiments are done with severely modified animals. These are lab strains of mice in small cages. All the lab strains have been selected to be tame and breed easily, which is exactly what the pheromone would tell them not to do.

Pfaff: All your studies have been with males. I guess Richard Axel has done something with females.

Dulac: Yes, maternal aggression.

Pfaff: Does the VNO distinguish novel from familiar odours?

Dulac: We have not tested that. So far our assays have been quite limited.

Keverne: It does learn, actually, in the context of the Bruce effect. The VNO/accessory olfactory system of the female learns on mating the identity of the male that mated.

Dulac: We couldn't test the Bruce effect in the TRPC2 knockout because our background is C57 and we are backcrossing into BALB/c.

C Blanchard: What are the downstream connections of V1/V2?

Dulac: Neurons expressing V1Rs and V2Rs send projections to different areas of the accessory olfactory bulb. Then the question is where the information is going from there. There have been two sets of experiments using dye injection that seem to suggest the two receptor populations actually project to the same areas in the amygdala. But the dye injection might not be precise enough to indicate what neuron type establishes synapses with anterior or posterior accessory bulb neurons. They might project to the same neurons. Our finding is that the mitral cells in the accessory olfactory bulb have very different properties. You could imagine that each channel detects different stimuli and processes them very differently. In turn, this will control the aggression and gender recognition. These are experiments that need to be done.

Skuse: What are the developmental consequences of being a TRPC2 knockout? If you have never been responsive to pheromones from day 0, this will have knock on effects for the development of other brain regions apart from those directly involved in pheromone detection.

Dulac: I have no data on this. There are likely to be compensatory changes. These would be quite fascinating to investigate. In particular, it is possible that the olfactory system is taking over part of the pathway.

Nelson: Are the knockout animals smaller than wild-type?

Dulac: No.

Skuse: Did you look anatomically at the amygdala to see if there were changes in the knockouts?

Dulac: We have looked briefly. This part of the amygdala hasn't really been studied properly. People have focused on the lateral amygdala.

Skuse: You had your control animals for comparison.

Dulac: We haven't looked carefully enough yet.

Keverne: It isn't entirely true that people haven't looked at the medial amygdala. There has been work done perfusing oxytocin into the medial amygdala, which has some effect on enabling familiarity. You can also get LTP in the medial amygdala if you stimulate the accessory olfactory bulb. There are also retrograde tracing studies, admittedly not from single neurons, but the evidence would suggest that the partitioning that occurs at the level of the AOB is not carried through to the medial amygdala.

Dulac: Has this been in the mouse?

Keverne: Yes.

Dulac: A lot of studies have been done in other species. In hamster it is clear that the VNO may be playing a different role.

Keverne: The oxytocin work is in the vole. In the rat, if you infuse β endorphin into the medial amygdala, the animal gets locked into investigative behaviour. It never engages mating behaviour.

Martinez: Is this system androgen-dependent in males?

Dulac: It is.

Robins: Does oestrogen control female pheromone synthesis?

Dulac: The female bit of this has never been studied properly, so I can't answer.

R *Blanchard:* We are more concerned with predatory odours. Have you examined these?

Dulac: No.

Craig: What happened to the VNO receptor family genes in humans? Do they have any function?

Dulac: Two or three genes have been found in the human genome that potentially have a functional open reading frame. This doesn't mean that their regulatory element if functional nor that they are actually expressed. My understanding is that if you look at a human VNO, there are no neurons that can be identified, so the receptors would have to be expressed somewhere else. They could be expressed in the main olfactory epithelium, but this wouldn't be very useful. We know that in the mouse and rat there are some 'lost' neurons in the olfactory epithelium, which express VNO receptors but which project nowhere. They might just be coming from progenitors that differentiate and die because they can't make any connections. The cluster of the M10 genes is deleted from the human genome.

Craig: That accounts for eight genes, but you were talking about 150 genes.

Dulac: They are all mutated. They probably started to be mutated at the same time as the split. It has been a long time. This doesn't mean that humans aren't sensitive to pheromones; their detection could occur through the main olfactory epithelium. We have data that even in the mouse this is occurring: there are some essential pheromones detected by the main olfactory epithelium.

Nelson: Has anyone looked at those genes early, prenatally, when humans do have a VNO?

Dulac: If they are mutated in the genome, they wouldn't be functional.

Pfaff: Your electrophysiological experiments were conducted with the receptor surfaces in the form of an open face. But the VNO is a long cul de sac. How do pheromones get out?

Dulac: It is a pump. When the animal is stimulated by novelty, the blood vessel close to it retracts and this expands the lumen. In contrast, when the blood vessel expands the lumen contracts.

Pfaff: It's a mixing function.

Brodkin: Are there specific receptors in the main olfactory epithelium that are involved in pheromone detection in the mouse? You implied this earlier.

Dulac: We have some data from our laboratory, but we don't know exactly what those receptors are. We think they are involved in the mating behaviour.

Koolhaas: Did I understand correctly that the TRPC protein is also involved in temperature perception?

Dulac: Not this specific one, but another member of the family: TRPV1 senses hot temperature, and TRPM8 is involved in cold detection. Each of the channel proteins has a distinct function and is expressed by different subsets of neurons.

Koolhaas: So your knockouts will not be impaired in temperature detection.

Keverne: When you were asking about the few VNO receptors found in the human genome, one possibility might be that they are expressed in testes.

Dulac: They are, but everything is expressed in testis.

Keverne: The reason I mention their presence on sperm is because the zona pellucida of the egg has a particular protein, a ZP protein. There are only three of these proteins in the genome: one of them is on the egg and another is in the VNO. It is worth looking to see whether such proteins might be concerned with sperm selection. The VNO is thought to be concerned with MHC and mate choice in mammals but, if we humans have lost this VNO system then perhaps we have moved over to a system where the selection is taking place at the level of a sperm fertilizing an egg.

Dulac: That is a fascinating hypothesis. As you know, olfactory receptors are also expressed in sperm.

Serotonergic gene inactivation in mice: models for anxiety *and* aggression?

Klaus-Peter Lesch

Clinical and Molecular Psychobiology, Department of Psychiatry and Psychotherapy, University of Wuerzburg, Fuechsleinstrasse 15, 97080 Wuerzburg, Germany

Abstract. Variation in genes coding for proteins that control serotonin (5-HT) system development, plasticity and function have been implicated in various aspects of complex behaviour including anxiety and aggression. Based on the remarkable progress in technologies that allow the alteration or elimination of individual genes to create transgenic animal models, gene knockout strategies further increase our knowledge about which serotonergic gene products are involved in behavioural traits. This overview selects anxiety and aggression as paradigmatic traits and behaviours, and focuses on mouse models which have been modified by deletion of genes coding for key players of serotonergic neurotransmission. In particular, phenotypic changes in mice bearing inactivation mutations of 5-HT_{1A} and 5-HT_{1B} receptors, 5-HT transporter, 5-HT neuron-specific transcription factor Pet1, monoamine oxidase A and genes related to 5-HT signalling will be discussed and major findings highlighted. However, because a missing gene might affect many developmental processes throughout ontogeny and compensatory mechanisms may be activated in knockouts, behavioural data from mice with targeted gene deletions should be interpreted with caution. The development of conditional knockout mice, in which a specific gene can be inactivated neurocircuit-specifically at any time, is therefore likely to avert the deficiencies associated with behavioural data from classical constitutive knockouts.

2005 Molecular mechanisms influencing aggressive behaviours. Wiley, Chichester (Novartis Foundation Symposium 268) p 111–146

In addition to genetic approaches, behavioural, functional imaging and pharmacological studies indicate that anatomically and functionally distinct neurocircuits as well as numerous neurotransmitters, growth factors, hormones and their intracellular signalling effectors influence fear, anxiety and avoidance, as well as fear–aggression interactions in humans and animal models. In humans, anxiety and aggression represent internal emotional states and are natural adaptive consequences of stress that help to cope with the stressor. However, unlike the relatively mild and brief anxiety responses resulting from a stressful event, anxiety disorders, such as generalized anxiety disorder, panic disorder and post-traumatic stress disorder are dysfunctional, chronic, persistent, and can get gradually worse unless treated.

Anxiety in rodents is defined as a high level of avoidance of novel and unfamiliar environments and increased fear reaction (Weiss et al 2000, Finn et al 2003). Other components such as autonomic activation, increased stress reactivity and neuroendocrine abnormalities are an integral part of anxiety-related behaviour. The design of an anxiety-related phenotype in mice partially or completely lacking a gene of interest during all stages of development (constitutive knockout) or in a spatiotemporal context (conditional knockout) is among the prime strategies directed at elucidating the role of genetic factors in fear and anxiety. Various approaches have been employed to detect and quantify 'anxiety-like' behaviours in mice and the majority postulate that aversive stimuli, such as novelty or potentially harmful environments induce a central state of fear and defensive reactions, which can be assessed and quantified through physiological and behavioural paradigms (Crawley & Paylor 1997, Crawley 1999). When rodents are introduced into a novel environment, they tend to move around the perimeter of the environment ('Open Field'). They stop occasionally and rear up, sniffing the walls and the floor. They initially spend very little time in the open centre of the area. If they have a choice, they will spend more time in a dark than in a brightly lit area ('Light–Dark Box'). Rodents will also spend more time in a small, elevated area enclosed by walls than in an elevated area without walls ('Elevated Plus Maze'). When they move from one delimited area into another, they often engage in a type of stretching-out behaviour. Anxiety-like behaviour often appears to contrast with exploratory behaviour indicating that avoidance and curiosity or novelty seeking are biologically related and may share common neurobehavioural systems (Fig. 2).

Investigations of the molecular mechanisms underlying aggressiveness (individual propensity for aggressive behaviour) and aggression (act of violence) seek to determine the neurobehavioural systems that underlie violence or other destructive behaviours in humans. Analogous to anxiety, anatomically and functionally distinct neural circuits influence aggression in animal models and humans (for review see Lesch 2003b, Lesch & Merschdorf 2000). Aggression is an umbrella term for various facets of agonistic behaviours which ranges from the establishment of hierarchies and dominance to antisocial behaviour and delinquency. In both humans and animals, the term aggression also comprises a spectrum of behaviours that are heterogeneous for neurobiological features and clinical phenomenology. Although most neurobiological studies of aggression and violence typically do not differentiate between defensive and offensive aggression, this distinction is likely relevant in understanding their genetic, functional neuroanatomical and neurochemical foundation.

It is believed that the propensity for impulsive aggression, which is relatively unplanned and spontaneous but often culminates in physical violence, is associated with a low threshold for activating negative affect (a combination of emotion and mood including anxiety, anger, distress and agitation) and with a failure to respond appropriately to the anticipated negative consequences of behaving aggressively. Social psychological research underscores the relation between emotion and aggression: negative affect such as fear and anxiety frequently precipitates, accentuates and modulates aggressive behaviour. While the impact of complex social and cultural variables on behaviour impedes simple extrapolation of animals to humans, clinical observation, experimental paradigms in the laboratory, and cluster/factor-analytic statistics have been used in attempts to subdivide aggressive behaviour. Based on different approaches human aggression may be differentiated into several subtypes depending on the presence or absence of causes or motivation, nature of trigger, characteristics of mediators, form of manifestation, direction and function. In consideration of the qualitatively distinct subtypes of aggression a dichotomy between an impulsive–reactive–hostile–affective subtype (defensive aggression) and a controlled–proactive–instrumental–predatory subtype (offensive aggression) has emerged as a promising construct (Fig. 1) (Vitiello & Stoff 1997).

Here, I focus on the nature of deficits in central serotonin (5-HT) system function that might predispose an individual to these aberrant, often deleterious or destructive forms of human and animal behaviour. Emphasis is also put on the molecular psychobiology of 5-HT in anxiety and aggression. Finally, an appraisal of behavioural and physiological consequences in mice with genetic manipulation of the serotonergic pathway is provided.

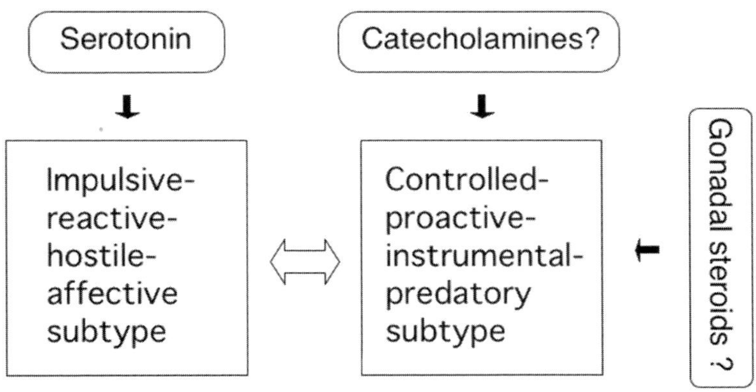

FIG. 1. In consideration of qualitatively distinct subtypes of aggression a dichotomy between an impulsive–reactive–hostile–affective subtype (defensive aggression) and a controlled–proactive–instrumental–predatory subtype (offensive aggression) has emerged as a promising construct. In addition to genetic approaches, functional imaging and pharmacological studies indicate that anatomically and functionally distinct neurocircuits as well as numerous neurotransmitters, growth factors, hormones and their intracellular signalling effectors influence fear–aggression interactions.

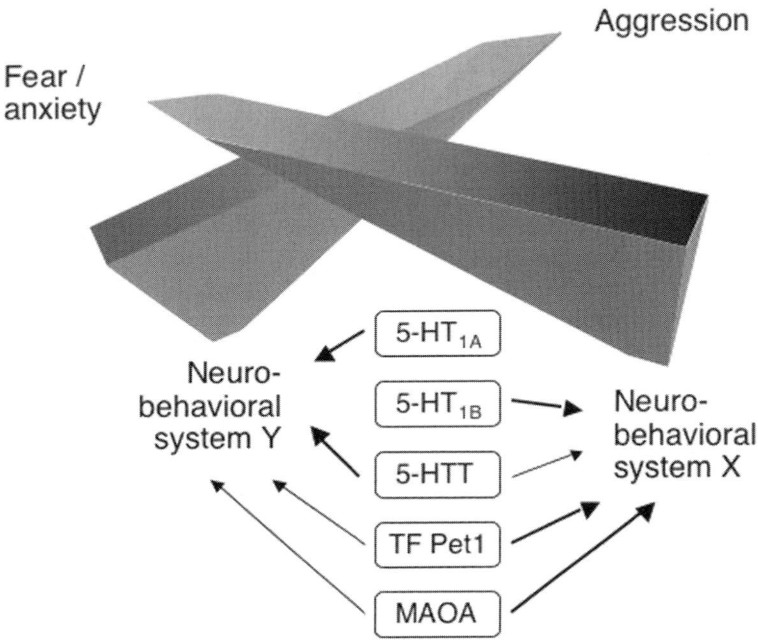

FIG. 2. Distinct gene products of the serotonergic pathway modulate different neurobehavioural systems which may be either synergistic or antagonistic.

Genetics of emotionality: quantitative trait locus and gene targeting strategies

Behaviours related to anxiety and aggression seem to delineate a biologically based model of dispositions to both normal and pathological functioning, with a continuum of genetic risk underlying behavioural dimensions that extend from normal to pathological. The documented heterogeneity of both genetic and environmental determinants suggests the inadequacy of searching for unitary causes. This vista has therefore increasingly encouraged the pursuit of dimensional approaches to behavioural genetics (Plomin et al 1994) and gene variants with a significant impact on the functionality of components of the 5-HT system are a rational strategy (Fig. 2).

While systematic studies of the patterns of inheritance of emotionality indicate that these traits are likely to be influenced by many genes (making it polygenic) or quantitative traits, behavioural genetics convincingly documents the significance of environmental factors. Individual differences in emotionality, and the ultimate behavioural consequences, such as fear, anxiety, avoidance, aggression, violence and self-injurious behaviour are relatively enduring, continuously distributed, as well as substantially heritable, and therefore are likely to result from additive or nonadditive interaction of genetic variations with environmental influences. Mouse strains that have been selectively bred to display a phenotype of interest are currently being used to identify genetic loci that contribute to behavioural traits. This quantitative trait loci (QTL) approach has been applied to a trait in mice distinctly called 'emotionality' (Flint et al 1995) and aggression (also see Brodkin 2005, this volume). But such linkage analyses provide only a rough chromosomal localization, whereas the next step, identifying the relevant genes by positional cloning, remains a challenging task.

While substantial similarities between human and murine avoidance, defence, aggression, or escape response exist, it remains obscure whether mice also experience subjective anxiety and associated cognitive processes similar to humans or whether defence responses or aggressiveness represent pathological expression of anxiety in humans. In general, pathological anxiety may reflect an inappropriate activation of normally adaptive, evolutionarily conserved defence reaction. It should therefore be practicable to elucidate both physiological and pathological anxiety by studying avoidant and defensive behaviour in mice using a broad range of anxiety and aggression paradigms to ensure comprehensive characterization of the behavioural phenotype. Since mice and humans share many orthologous genes mapped to syntenic chromosomal regions, it is conceivable that individual genes successfully identified for one or more types of murine anxious or aggressive behaviour may be developed as animal models for human anxiety and aggression, respectively. Although mice and humans species share >90% of their genes in common and the underlying molecular and neurobiological mechanisms found in anxious or aggressive mice are also reported in humans displaying anxiety or aggression, 'emotional' behaviour observed in mice is rarely directly comparable. Following chromosomal mapping of polymorphic genes and evaluation of gene function using knockout (KO) mutants, behavioural parameters, including the type of emotionality, measure of emotionality, test situation and opponent type are investigated (Maxson 1996). Thus, the combination of elaborate genetic and behavioural analyses provides an attractive tool to discover new candidate genes with effects on variation and development of one or more forms of murine anxiety- and aggression-related behaviour.

Notwithstanding the confounding issues, the KO mouse remains a powerful tool for modelling the genetic basis of behaviour. Constitutively created mutations mimic genetic variability in the sense that they are present during the entire developmental process. In contrast, the spectrum seen in behaviour is the result of developmental adaptation. As demonstrated in KO mice, the developmental impact of an inactivated gene might be more prominent than the actual loss of its function that occurs during adulthood. A major challenge is therefore the identification of neurogenetic mechanisms that underlie emotionality. The obvious difficulties in the attempts to unravel the genetic basis of emotion regulation also reflect on the complex nature of these traits, which is expressed in many different facets. Such facets can be distinguished by specific testing procedures that identify particular categories of fear, anxiety, and avoidance as well as of aggression, such as anxiety-driven defensive aggression, shock-induced or irritability-associated aggression, isolation-induced offensive aggression, predatory aggression, maternal aggression and infanticide (Tecott & Barondes 1996).

Despite the usefulness of the gene KO strategy in identifying specific gene products that may be involved in anxiety or aggression, this approach is limited to known candidate genes. Due to the complexity in the expression of emotional behaviour, it is impossible to predict which genes contribute to the variability of this trait in different populations. Thus, QTL analysis, although technically demanding, should ultimately prove to be an important complementary approach, because it is likely that identification of the particular alleles of the various genes that influence emotionality in inbred strains will facilitate elucidation of epistatic (gene×gene) interactions as well as the phenomenon of pleiotropy, the multiple and apparently independent effects of a genes on phenotypical expression.

Serotonergic system in emotion regulation

A neural circuit composed of several regions of the prefrontal cortex, amygdala, hippocampus, medial preoptic area, hypothalamus, anterior cingulate cortex, ventral striatum and other interconnected structures, and involving multiple neurotransmitter systems such as 5-HT has been implicated in emotion regulation including the associated negative affect of fear, anxiety anger, affect and agonistic behaviour (for review see Davidson et al 2000). Although the brain systems mediating anxiety and aggression appear to be rather conserved among mammals, several details of the regulatory pathways are likely to be species specific. While both genetic and environmental factors contribute to the structure and function of this circuitry, the amygdala is central to processes of learning to associate stimuli with events that are either punishing or rewarding. Anxiety- and aggression-related circuits involve pathways transmitting information to and from the amygdala to various neural networks that control the expression of avoidant, defensive or aggressive reactions, including behavioural, autonomic and stress hormone responses. While pathways from the thalamus and cortex (sensory and prefrontal) project to the amygdala, inputs are processed within intra-amygdaloid circuitries and outputs are directed to the hippocampus, brain stem, hypothalamus and other regions.

The function of the amygdala in emotion regulation is highly complex. Perception of danger or threat are transmitted to the lateral nucleus of the amygdala, which projects to the basal nuclei where information regarding the social context derived from orbitofrontal projections is integrated with the perceptual information. Behavioural responses can then be initiated via activation of projections from the basal nuclei to various association cortices, while physiological responses can be produced via projections from the basal nuclei to the central nucleus and then to the hypothalamus and brainstem. Excessive or insufficient activation of the amygdaloid complex leads to either disproportionate negative affect or impaired sensitivity to social signals. The orbitofrontal cortex, through its connections with other domains of the prefrontal cortex and with the amygdala, plays a critical role in limiting impulsive outbursts, and the anterior cingulate cortex recruits other neural systems during fear- or anger-evoked arousal and other negative emotions. Functional or structural abnormalities in these regions or in their interconnections can modify negative affect and aggressiveness and it has therefore been suggested that impulsive aggression and violence arise as a consequence of a deficiency in emotion regulation (Davidson et al 2000).

In humans, non-human primates, and other mammals, preclinical and clinical studies have accumulated substantial evidence that the serotonergic signalling pathway is a major modulator of emotional behaviour including fear and anxiety, as well as impulsivity and aggression, and integrates complex brain functions such as cognition, sensory processing and motor activity. This diversity of these functions is due to the fact that 5-HT orchestrates the activity and interaction of several other neurotransmitter systems. The central 5-HT system, which originates in the midbrain and brainstem raphe complex, is widely distributed throughout the brain and its chemical messenger is viewed as a master control neurotransmitter within this highly elaborate system of neural communication mediated by 14+ pre- and postsynaptic receptor subtypes with a multitude of isoforms (e.g. functionally relevant splice variants) and subunits. The prefrontal cortex receives major serotonergic input, which appears dysfunctional in individuals who are impulsively violent. Individuals vulnerable to faulty regulation of negative emotion are therefore at risk for aggression and violence.

The level of 5-HT in the synaptic (and extrasynaptic) space is restricted by the synchronized action of at least three components. Firing of raphe 5-HT neurons is controlled by 5-HT_{1A} autoreceptors located in the somatodendritic section of neurons. Release of 5-HT at the terminal fields is regulated by the 5-HT_{1B} receptor. Once released, 5-HT is taken up by the 5-HT transporter located at the terminals (as well as the somatodendritic fraction) of 5-HT neurons, where it is eventually metabolized by monoamine oxidase A. The action of 5-HT as a messenger is tightly regulated by its synthesizing and metabolizing enzymes, and the 5-HT transporter. It is therefore likely that modification or removal of one of these components affects extracellular levels of 5-HT.

Serotonergic raphe neurons diffusely project to all brain regions implicated in aggressive behaviour, while neurons in anxiety- and aggression-mediating areas are rich in both 5-HT_1 and 5-HT_2 receptor subtypes. In addition to its role as a neurotransmitter, 5-HT is an important regulator of morphogenetic activities during early brain development as well as during adult neurogenesis and plasticity, including cell proliferation, migration, differentiation and synaptogenesis (Azmitia & Whitaker-Azmitia 1997, Lauder 1993, Di Pino et al 2005, Gaspar et al 2003).

Genes of the serotonergic pathway

Serotonin receptors

Fourteen different 5-HT receptor subtypes confer the effects of 5-HT upon neuronal or other cells. Pharmacological classification based on studies of ligand binding to receptor subtypes and of signal transduction pathway responses to agonists/antagonists were traditionally employed to delineate four 5-HT receptor subfamilies, $5\text{-HT}_{1\text{–}4}$. Gene identification efforts have eventually not only validated this classification but also uncovered the existence of several novel 5-HT receptor subtypes ($5\text{-HT}_{1E/F}$, $5\text{-HT}_{3A/B}$, $5\text{-HT}_{5A/B}$, 5-HT_6 and 5-HT_7) (Barnes & Sharp 1999, Hoyer & Martin 1997). In $5\text{-HT}_{2\text{–}7}$ receptor genes the coding region is interrupted by introns, whereas the genes for $5\text{-HT}_{1A\text{–}F}$ receptors contain no intronic sequences. The 5-HT_{2B}, 5-HT_4, 5-HT_6 and 5-HT_7 receptors are alternatively spliced and RNA editing of the 5-HT_{2C} receptor subtype in the second intracellular loop has been reported to confer differential receptor properties. A 5-HT_{1A} receptor variant, Gly22Ser, shows differences in agonist-induced down-regulation compared with the wild-type 5-HT_{1A} receptor allele, thus increasing the complexity of naturally occurring 5-HT receptor structural variants. The challenge now is to identify the physiological impact of these gene products, establish their specific functionality with respect to distinct neurocircuits of emotion regulation, design selective agonists/antagonists, and determine potential therapeutic application of these novel compounds.

The molecular characterization of different 5-HT receptor subtypes has simplified the elucidation of gene transcription, mRNA processing, and translation as well as intracellular trafficking and posttranslational modification relevant to synaptic and postreceptor signalling. Transcriptional control regions have been cloned for several 5-HT receptor subtypes and functional promoter mapping data are available for the 5-HT_{1A}, 5-HT_{2A}, 5-HT_{2C} and 5-HT_3 receptor genes.

Serotonin 1A receptor. The 5-HT_{1A} receptor subtype has long been implicated in the pathophysiology of anxiety and depression; its role as a molecular target of anxiolytic and antidepressant drugs is also well established (Griebel 1995, 2000, Olivier et al 1999). Patients with panic disorder and depression display an attenuation of 5-HT_{1A} receptor-mediated hypothermic and neuroendocrine responses, reflecting a reduced responsivity of both pre- and postsynaptic 5-HT_{1A} receptors (Lesch et al 1990a, 1992). Likewise, a decrease in 5-HT_{1A} ligand binding has been shown in postmortem brain of depressed suicide victims (Cheetham et al 1990) as well as in forebrain areas such as the medial temporal lobe and in the raphe of depressed patients elicited by positron emission tomography (PET) (Sargent et al 2000, Drevets et al 1999). Both glucocorticoid administration and chronic stress, a pathogenetic factor in affective disorders, have also been demonstrated to result in down-regulation of 5-HT_{1A} receptors in the hippocampus in animals (Wissink et al 2000, Flügge 1995, Lopez et al 1998). While deficits in hippocampal 5-HT_{1A} receptor function may contribute to the cognitive abnormalities associated with affective disorders, recent work suggests that activation of this receptors stimulates neurogenesis in the dentate gyrus of the hippocampus. By using both a mouse model with a targeted ablation of the 5-HT_{1A} receptor and radiological methods, Santarelli and coworkers (Santarelli et al 2003) have provided persuasive evidence that 5-HT_{1A}-activated hippocampal neurogenesis is essentially required for the behavioural effects of long-term antidepressant treatment with 5-HT reuptake inhibitors.

Somewhat unexpectedly, down-regulation and hyporesponsivity of 5-HT_{1A} receptors in patients with major depression are not reversed by antidepressant drug treatment (Lesch et al 1990b, 1991, Sargent et al 2000), raising the possibility that low receptor function is a trait feature and therefore a pathogenetic mechanism of the disease. In line with this notion, evidence is accumulating that a polymorphism in the transcriptional control region of the 5-HT_{1A} receptor gene (*HTR1A*) resulting in allelic variation of 5-HT_{1A} receptor expression, is associated with personality traits of negative emotionality including anxiety and depression (Neuroticism and Harm avoidance) (Strobel et al 2003; Table 1) as well as major depressive disorder, suicidality and panic disorder (Lemonde et al 2003, Rothe et al 2004).

5-HT_{1A} receptors operate both as somatodendritic autoreceptors and as postsynaptic receptors. Somatodendritic 5-HT_{1A} autoreceptors are predominantly located on serotonergic neurons and dendrites in the brainstem raphe complex and their activation by 5-HT or 5-HT_{1A} agonists decreases the firing rate of serotonergic neurons and subsequently reduces the synthesis, turnover, and release of 5-HT from nerve terminals in projection areas. Postsynaptic 5-HT_{1A} receptors are widely distributed in forebrain regions that receive serotonergic input, notably in the cortex, hippocampus, septum, amygdala and hypothalamus. Their activation results in membrane hyperpolarization and decreased neuronal excitability. Hippocampal heteroreceptors mediate neuronal inhibition by coupling to GIRK2 potassium channels. Physiological responses depend upon the function of the target cells (e.g. hypothermia, activation of the hypothalamic pituitary adrenocortical system). Moreover, 5-HT_{1A} receptor expression is modulated by steroid hormones and 5-HT_{1A}-mediated signalling is an important regulator of gene expression through its coupling to G proteins that inhibit adenylyl cyclase and through modulation of GIRK2 channels.

The effects of 5-HT_{1A} receptor selective agents, such as the agonist 8-hydroxy-2-(di-n-propylamino) tetralin (8-OH-DPAT), and the partial agonists ipsapirone and gepirone, have been extensively studied in rodents (De Vry 1995). Both agonists and partial agonists induce a dose-dependent anxiolytic effect, which correlates with the inhibition of serotonergic neuron firing, decrease of 5-HT release, as well as the reduction of 5-HT signalling at postsynaptic target receptors. Blockade of the negative feedback by selective 5-HT_{1A} receptor antagonists, such as WAY 100635 increases firing of the serotonergic neurons but exerts no effect on 5-HT neurotransmission or behaviour (Olivier & Miczek 1999), while the combination with selective 5-HT reuptake inhibitors augments increases in 5-HT levels in terminal regions.

The converging lines of evidence that receptor deficiency or dysfunction is involved in mood and anxiety disorders encouraged investigators to genetically manipulate the 5-HT_{1A} receptor in mice (Heisler et al 1998, Parks et al 1998, Ramboz et al 1998). Mice with a targeted inactivation of the *Htr1a* show a complete lack of ligand binding to brain 5-HT_{1A} receptors in *Htr1a* (−/−) KO mice, with intermediate binding in the heterozygous (+/−) mice. Importantly, a similar behavioural phenotype characterized by increased anxiety-related behaviour and stress reactivity in several avoidance and behavioural despair paradigms was observed in three different *Htr1a* KO mouse strains (Lesch & Mössner 1999).

Htr1a KO mice consistently display a spontaneous phenotype that is associated with a gender-modulated and gene-dose-dependent increase of anxiety-related behaviours (Heisler et al 1998, Parks et al 1998, Ramboz et al 1998). With the exception of an enhanced sensitivity of terminal 5-HT_{1B} receptors, no major neuroadaptational changes were detected. Worthy of note is that this behavioural phenotype was observed in animals in which the mutation was bred into mice of Swiss-Webster (SW), C57BL/6J, and 129/SV backgrounds, substantiating the assumption that this behaviour is an authentic consequence of reduced or absent

TABLE 1 Gene products of the serotonergic pathway involved in anxiety and aggression

	Mice		*Humans*	
	Fear/anxiety	*Aggression*	*Anxiety*	*Aggression*
5-HT receptors				
1A	↑	—	+	—
1B	—	↑	—	+
5-HT transporter	↑	↓	+	+
TF Pet1	↑	↑	nd	nd
Monoamine oxidase A	—	↑	+	+

↑/↓, increase/decrease in anxiety and/or aggression.
+, association or linkage with anxiety and/or aggression.
−, no effect.
nd, not determined.

5-HT_{1A} receptors. While all investigators used Open Field exploratory behaviour as a model for assessing anxiety, two groups confirmed that *Htr1a* KO mice had increased anxiety by using other models, the Elevated Zero Maze or Elevated Plus Maze test (Heisler et al 1998, Ramboz et al 1998; Table 1). These ethologically based conflict models test fear and anxiety-related behaviours based on the natural tendencies for rodents to prefer enclosed, dark spaces versus their interest in exploring novel environments.

Activation of presynaptic 5-HT_{1A} receptors provide the brain with an autoinhibitory feedback system controlling 5-HT neurotransmission. Thus, enhanced anxiety-related behaviour most likely represents a consequence of increased terminal 5-HT availability resulting from the lack or reduction in presynaptic somatodendritic 5-HT_{1A} autoreceptor negative feedback function (Lesch & Mössner 1999). Although extracellular 5-HT concentrations and 5-HT turnover appear to be unchanged in the brain of *Htr1a* KO mice on the SW and 129/SV backgrounds, indirect evidence for increased presynaptic serotonergic activity resulting in elevated synaptic 5-HT concentrations is provided by the compensatory up-regulation of terminal 5-HT release-inhibiting 5-HT_{1B} receptors (Olivier et al 2001, Toth 2003). In contrast to *Htr1a* KO mice with a SW or 129/SV background, extracellular 5-HT concentrations were significantly elevated in mutant C57Bl/6 mice in the frontal cortex ad hippocampus (Parsons et al 2001). This may reflect a lack in compensatory changes in 5-HT_{1B} receptor and is consistent with findings that C57Bl/6 mice are more aggressive and susceptible to drugs of abuse than many other strains.

Several studies addressed electrophysiological properties of both presynaptic serotonergic neurons and postsynaptic hippocampal neurons *Htr1a* KO mice. A robust increase in the mean firing rate in dorsal raphe neurons was also reported, although a considerable number of neurons was firing in their normal range and 5-HT release was not altered (Richer et al 2002). Moreover, mutant mice showed an absence of paired-pulse inhibition in the CA1 region and lack of paired-pulse facilitation in the dentate gyrus, suggesting altered hippocampal excitability and impaired plasticity of the hippocampal network with consequence for cognition, learning and memory (Sibille et al 2000).

This mechanism is also consistent with models of fear and anxiety that are primarily based upon pharmacologically derived data. The cumulative reduction in serotonergic impulse flow to septohippocampal and other limbic and cortical areas involved in the control of anxiety is believed to explain the anxiolytic effects of ligands with selective affinity for the 5-HT_{1A} receptor in some animal models of anxiety-related behaviour. This notion is based, in part, on evidence that 5-HT_{1A} agonists (e.g. 8-OH-DPAT) and antagonists (e.g. WAY 100635) have anxiolytic or anxiogenic effects, respectively. However, to complicate matters further, 8-OH-DPAT has anxiolytic effects when injected in the raphe nucleus, whereas it is anxiogenic when applied to the hippocampus. Thus, stimulation of postsynaptic 5-HT_{1A} receptors has been proposed to elicit anxiogenic effects, while activation of 5-HT_{1A} autoreceptors is thought to induce anxiolytic effects via suppression of serotonergic neuronal firing resulting in attenuated 5-HT release in limbic terminal fields.

Since the 5-HT_{1A} receptor is expressed in different brain subsystems, it is of interest to clarify whether pre- or postsynaptic receptors are required to maintain normal expression of anxiety-related behaviour in mice. With an elegant conditional rescue approach, Gross et al (2002) illustrated that expression of the 5-HT_{1A} receptor in the hippocampus and cortex but not in the raphe nuclei is required to rescue the behavioural phenotype in *Htr1a* KO mice. The findings indicate that deletion of the 5-HT_{1A} receptor in mice, specifically in forebrain structures, results in a robust anxiety-related phenotype and that this phenotype in *Htr1a* mice is caused by the absence of the receptor during a critical period of postnatal development, whereas inactivation of 5-HT_{1A} in adulthood does not affect anxiety. Even more importantly, the findings further support the notion of a central role for 5-HT in the early development of neurocircuits mediating emotion (Lesch 2003a, Di Pino et al 2005). Although there is converging evidence that the 5-HT_{1A} receptor mediates anxiety-related behaviour, the neurodevelopmental mechanism that renders *Htr1a* KO mice more anxious is highly complex and remain to be elucidated in its details.

While increased 5-HT availability and activation of other serotonergic receptor subtypes that have been shown to mediate anxiety (e.g. 5-HT_{2C} receptor) may contribute to increased anxiety in rodent models, multiple downstream neurotransmitter pathways or neurocircuits, including GABAergic, noradrenergic, glutamatergic and peptidergic transmission, as suggested by overexpression or targeted inactivation of critical genes within these systems (Lesch 2001), have been implicated to participate in the processing of this complex behavioural trait. Since avoidance induced by conflict and fear is only one dimension of anxiety-related responses; other components including autonomic systems activation, responsiveness to stress, 5-HT dynamics and neuronal excitability in limbic circuitries appear to be involved in fear and anxiety.

As one facet of anxiety-like behaviour, *Htr1a* KO mice show genotype-dependent and background strain-unrelated increase in stress reactivity in two paradigms of behavioural despair, the Forced Swim and Tail Suspension tests (Heisler et al 1998, Parks et al 1998, Ramboz et al 1998). Autonomic manifestations of anxiety and stress responsiveness in a novel environment or when exposed to other stressors increased heart rate and body temperature as well as attenuated release of corticosterone is also a characteristic of *Htr1a* KO mice (Groenink et al 2003). The reduced immobility in stress/antidepressant test models is either due to an increased serotonergic tone resulting from the compromised 5-HT_{1A} autoreceptor-dependent negative feedback regulation or enhanced dopamine and norepinephrine function because it is reversed by pretreatment with α-methyl-*para*-tyrosine, but not by *para*-chlorophenylalanine (Mayorga et al 2001).

Although the behaviour of *Htr1a* KO mice in various stress-related paradigms is more consistent with increased emotionality, their behaviour essentially corresponds with the performance of rodents treated with antidepressants. The role of 5-HT_{1A} receptors in the therapeutic action of antidepressant drugs has attracted extraordinary interest; however, there is substantial conflicting evidence regarding the involvement of other serotonergic receptor subtypes and neurotransmitter systems or neurocircuits that interact with 5-HT neurotransmission. Electrophysiological studies in rats indicate that each class of antidepressants enhances 5-HT neurotransmission via differential adaptive changes in the 5-HT_{1A} receptor-modulated negative feedback regulation that eventually leads to an overall increase of terminal 5-HT (for review see Blier & de Montigny 1998) and desensitization of 5-HT_{1A} responsivity following antidepressant treatment has been demonstrated in rodents and humans. While the neuroadaptive mechanism of antidepressant action of tricyclics or selective 5-HT reuptake inhibitors is exceedingly complex, as the onset of clinical improvement commonly takes 2–3 weeks or more after initiation of antidepressant drug administration, progressive functional desensitization of pre- and postsynaptic serotonergic receptors, including the 5-HT_{1A}, 5-HT_{1B} and 5-HT_{2A}, that is set off by blockade of the 5-HT transporter, has been implicated in these delayed therapeutic effects. In conclusion, the phenotypic similarity between anxiety-related behaviour and stress reactivity in humans and *Htr1a* KO mice powerfully validates the practicability of KO animal models.

Serotonin 1B receptor. The 5-HT_{1B} receptor has long been implicated in various models of rodent agonistic behaviour which differentiate between defensive/flight and offensive aggression models have been described (Olivier & Mos 1992, Olivier et al 1995).

Nevertheless, documentation of the contribution of other 5-HT receptor subtypes in aggression is still constrained by the lack of selective ligands, but drugs that target the 5-HT_{2A}, 5-HT_{2C} or 5-HT_3 sites have generally not influenced aggressiveness. It may thus currently be more instructive to evaluate aggressive behaviour systematically in KO mice of various 5-HT receptor subtypes. With the identification of numerous other 5-HT receptor subtypes, much work remains to be done to clarify the specific role of these 5-HT receptors to discriminate between pre- and postsynaptic effects on aggression and the interactions among 5-HT receptor subtypes that underlie aggression.

5-HT_{1B} receptors are expressed in the basal ganglia, central grey, lateral septum, hippocampus, amygdala and raphe nuclei. They are located predominantly at presynaptic terminals inhibiting 5-HT release or, as heteroreceptors, modulating the release of other neurotransmitters. Selective agonists and antagonists for 5-HT_{1B} receptors are largely lacking, but indirect pharmacological evidence suggests that 5-HT_{1B} activation influences food intake, sexual activity, locomotion and emotionality including particularly impulsivity and aggression.

The 5-HT_{1B} receptor was the first subtype to have its gene inactivated by classical homologous recombination (Saudou et al 1994). Mice with a targeted disruption of the 5-HT_{1B} gene (*Htr1b*) therefore facilitated investigation of the concept of 5-HT-related impulsivity in the context of aggressive behaviour. Two of the behaviours, locomotion and aggression, postulated to be modulated by 5-HT_{1B} receptors were analysed (Ramboz et al 1996). Wild-type and *Htr1b* (−/−) KO mice were found to display similar levels of locomotor activity in an open field. Impulsivity and aggression-related behaviour of $Htr1b^{-/-}$ male mice was assessed by isolation and subsequent exposure to a non-isolated male wild-type intruder mouse. The latency and number of attacks displayed by the KO mice were used as indices of aggression. $Htr1b^{-/-}$ mice, when compared with wild-type mice, showed more rapid, more intense and more frequent attacks. Lactating female $Htr1b^{-/-}$ mice also attack unfamiliar male mice more rapidly and violently. In addition to increased aggression, KO mice acquire cocaine self-administration faster and ingest more ethanol than controls (Brunner & Hen 1997). Thus, the 5-HT_{1B} receptor modulates not only motor impulsivity and aggression but also addictive behaviour.

These results further support the notion that the 5-HT_{1B} subtype is not the only 5-HT receptor regulating aggression and that distinct receptor subtypes modulate different dimensions of behaviour which may be either synergistic or antagonistic (Fig. 2). Opposite to *Htr1a* KO mice (also see section 4.1.1.), *Htr1b* KO mice are more reactive and more aggressive but show dramatically less anxiety-related behaviour than control mice, although both 5-HT_{1A} and 5-HT_{1B} receptors control the tone of the serotonergic system and mediate some of the postsynaptic 5-HT effects (Zhuang et al 1999). The regional variation of 5-HT receptor expression and the complex autoregulatory processes of 5-HT function which are operational in different brain areas may lead to a plausible hypothesis to explain this apparent contradiction. Assessment of 5-HT_{1A} receptor expression in male mice selected for high and low offensive aggression showed that high-aggressive mice, characterized by a short attack latency, decreased plasma corticosterone concentration and increased levels of 5-HT_{1A} mRNA in the dorsal hippocampus (dentate gyrus and CA1) compared to low-aggressive mice that had long attack latency and high plasma corticosterone levels (Korte et al 1996). Increased postsynaptic 5-HT_{1A} receptor radioligand binding was found in dentate gyrus, CA1, lateral septum and frontal cortex, whereas no difference in ligand binding was found for the 5-HT_{1A} autoreceptor on cell bodies in the dorsal raphe nucleus. These results suggest that high offensive aggression is associated with reduced (circadian peak) plasma corticosterone and increased postsynaptic 5-HT_{1A} receptor availability in limbic and cortical regions.

At the next level of complexity signalling through 5-HT receptors involves different transduction pathways, and each receptor subtype modulates distinct, though frequently interacting, second messenger systems and multiple effectors. The discovery of a considerable number of hyperaggressive mutant strains in the course of gene KO experiments highlights the extraordinary diversity of genes involved in the genetic influence on impulsivity and aggression. Interestingly, genetic support for a role of 5-HT in aggression also derives from mice lacking specific genes, including neuronal nitric oxide synthase (nNOS) and neural cell adhesion molecule (NCAM), that either directly or indirectly affect 5-HT turnover or 5-HT receptor sensitivity. Male *nNos* KO mice and wild-type mice in which nNOS is pharmacologically suppressed are highly aggressive (Chiavegatto et al 2001). Excessive aggression of *nNos* KO mice is caused by a selective decrease in 5-HT turnover and deficient 5-HT_{1A} and 5-HT_{1B} receptor function in brain regions regulating emotion. The neural cell adhesion molecule (NCAM) plays a critical role during brain development and in adult plasticity, and *Ncam* KO mice display elevated anxiety and aggression (Stork et al 1999). Lower doses of 5-HT_{1A} agonists are necessary to reduce anxiety and aggressiveness in the *Ncam* KO mice, suggesting a functional change in the 5-HT_{1A} receptor, although 5-HT_{1A} binding as well as brain 5-HT and 5-hydroxyindolacetic acid (5-HIAA) tissue concentrations were unaltered. Taken together, these findings indicate an involvement of nNOS and NCAM and the 5-HT system through 5-HT_{1A} and 5-HT_{1B} receptors, but the specific molecular mechanisms in aggression remain to be elucidated.

Serotonin transporter. High-affinity 5-HT transport into the presynaptic neuron is mediated by a single protein, the 5-HT transporter (5-HTT, SERT), which is regarded as initial sites of action of antidepressant drugs and several neurotoxic compounds. Tricyclic antidepressants, such as prototypical imipramine, and the selective 5-HT uptake inhibitors, paroxetine, citalopram and sertraline, occupy several pharmacologically distinct sites overlapping at least partially the substrate binding site and are widely used in the treatment of depression, anxiety and impulse control disorders, as well as substance abuse including alcoholism.

While in adult brain 5-HTT expression appears to be restricted to raphe neurons, it has been detected in the sensory areas of the cortex and thalamus during perinatal development (Lesch & Murphy 2003). Cloning of the *5HTT* gene (*HTT, SLC6A4*) has identified a protein with 12 transmembrane domains and studies using site-directed mutagenesis and deletion mutants indicate that distinct amino acid residues participate in substrate translocation and competitive antagonist binding. 5-HTT function is acutely modulated by post-translational modification. Moreover, several intracellular signal transduction pathways converge on the transcriptional apparatus of the *HTT* regulating its expression. A polymorphism in the transcriptional control region of the human *HTT* gene that results in allelic variation in functional 5-HTT expression is associated with anxiety, depression, and aggression-related personality traits (Lesch et al 1996, Lesch 2003a). In addition to the exploration of the impact of allelic variation in 5-HTT expression on anxiety, depression and aggression-related personality traits, a role of the regulatory and structural *HTT* variation has been suggested in a variety of diseases such as depression, bipolar disorder, anxiety disorders, eating disorders, substance abuse, autism, schizophrenia and neurodegenerative disorders (Lesch & Murphy 2003).

Anxiety- and aggression-related behaviour

The converging evidence that 5-HTT deficiency plays a role in anxiety and related disorders lead to the generation of mice with a targeted inactivation of the *Htt* (*Slc6a4*). Behaviour of the *Htt* KO mice was tested in a variety of conditions evaluating fear, avoidance, conflict, stress responsiveness, status of the neuroendocrine system and effects of various pharmacological agents on the behaviour. In particular, anxiety-related behaviours were characterized using a battery of tests including Open Field, Elevated Plus Maze and Light–Dark Box.

In these tests both male and female *Htt* KO mice show consistently increased anxiety-like behaviour and inhibited exploratory locomotion. The selective 5-HT_{1A} receptor antagonist WAY 100635 produced an anxiolytic effect in the elevated plus-maze in *Htt* KO mice suggesting that the abnormalities in anxiety-like and exploratory behaviour is mediated by the 5-HT_{1A} receptor (Holmes et al 2003). Unlike heterozygous $Htr1a^{+/-}$ mice, $Htt^{+/-}$ mice, in which transporter binding sites are reduced by approximately 50%, were similar to controls on most measures of anxiety-like behaviour. However, changes in exploratory behaviour in $Htt^{+/-}$ mice were limited to specific measures under baseline conditions, but extended to additional measures under more stressful test conditions. This observation is in accordance with reduced aggressive behaviour in $Htt^{+/-}$ mice that is limited to specific measures and test conditions. While male $Htt^{-/-}$ mice are slower to attack the intruder and attacked with less frequency than control littermates, heterozygous $5Htt^{+/-}$ mice were as quick to attack, but made fewer overall attacks, as compared to wild-type controls. Aggression increased with repeated exposure to an intruder in $5Htt^{+/-}$ and control mice, but not in $Htt^{-/-}$ mice.

These behavioural alterations in $Htt^{+/-}$ mice are paralleled by robust perturbations in serotonergic homeostasis that are intermediate between $-/-$ mice and controls in a gene-dose-dependent manner including elevated extracellular 5-HT, decreased 5-HT neuron firing in the dorsal raphe, and reduced 5-HT_{1A} receptor expression and function (Li et al 2000, 1999, Gobbi et al 2001). The evidence that serotonergic dysfunction in $Htt^{+/-}$ mice may manifest and become noticeable as behavioural abnormalities only under challenging environmental conditions strongly support the disposition-stress model of affective and anxiety disorders (Murphy et al 2003).

Analogous to *Htr1a* KO mice, the neural mechanisms underlying increased anxiety-related behaviour and reduced exploratory locomotion in mice with a disruption of the *Htt* may relate to excess serotonergic neurotransmission which is expected to cause enhanced activation of postsynaptic 5-HT receptors. Both *in vivo* microdialysis in striatum and *in vivo* chronoamperometry in hippocampus revealed that *Htt* KO mice exhibit an approximately five- to ninefold increase in extracellular concentrations of 5-HT and an absence of transporter-mediated clearance, although brain tissue 5-HT concentrations are markedly reduced by 40–60% (Bengel et al 1998).

Excess of extracellular 5-HT activates the negative autoinhibitory feedback and reduces cellular 5-HT availability by stimulating 5-HT_{1A} receptors, which results in their desensitization and down-regulation in the midbrain raphe complex and, to a lesser extent, in hypothalamus, septum and amygdala but not in the frontal and hippocampus (Li et al 2000). Although postsynaptic 5-HT_{1A} receptors appear to be unchanged in frontal cortex and hippocampus, indirect evidence for decreased presynaptic serotonergic activity but reduced 5-HT clearance resulting in elevated synaptic 5-HT concentrations is provided by compensatory alterations in 5-HT synthesis and turnover, down-regulation of terminal 5-HT release-inhibiting 5-HT_{1B} receptors (Fabre et al 2000).

Therefore, a partial down-regulation of postsynaptic 5-HT_{1A} receptors in some forebrain regions but a several-fold increase in extracellular concentrations of 5-HT in *Htt* KO mice could still cause excess net activation of postsynaptic 5-HT_{1A} receptors, resulting in increased anxiety-like behaviour, and its reversal by WAY 100635 (Holmes et al 2003). However, administration of WAY 100635 antagonizes not only postsynaptic 5-HT_{1A} receptors in forebrain regions but also acts at somatodendritic autoreceptors in the raphe nuclei and electrophysiological studies show that WAY 100635 causes a reversal of markedly reduced spontaneous firing rates of 5-HT neurons in the dorsal raphe dorsal nucleus of $Htt^{-/-}$ mice, indicating that the net effect of WAY 100635 on serotonergic neurotransmission in *Htt* KO mice may be more complex than anticipated (Gobbi et al 2001).

Taken together, these findings add to an emerging picture of abnormalities in *Htt* KO mice across a range of behavioural, neuroendocrine and physiological parameters associated with emotional disorders, including marked increases in adrenocorticotropin (ACTH) concentrations in responses to stress (Li et al 1999), increased sensitivity to drugs of abuse such as cocaine (Sora et al 1998, 2001), altered gastrointestinal motility (Chen et al 2001) and disturbed REM sleep (Wisor et al 2003). Finally, given the absence of the 5-HTT throughout ontogeny, *Htt* KO mice also provide a research tool for studying the potential for neurodevelopment abnormalities affecting anxiety-like behaviour.

Analogous to the situation in *Maoa* KO mice (also see section on *Monoamine oxidase A*, p 131), inactivation of the *Htt* profoundly disturbs formation of the somatosensory cortex (SSC) with altered cytoarchitecture of cortical layer IV, the layer that contains synapses between thalamocortical terminals and their postsynaptic target neurons (Persico et al 2001). Brains of *Htt* KO mice display no or only very few barrels. Cell bodies as well as terminals, typically more dense in barrel septa, appear homogeneously distributed in layer IV of adult *Htt* KO brains. Injections of a 5-HT synthesis inhibitor within a narrow time window of 2 days postnatally completely rescued formation of SSC barrel fields. Of note, heterozygous KO mice develop all SSC barrel fields, but frequently present irregularly-shaped barrels and less defined cell gradients between septa and barrel hollows. These findings demonstrate that excessive concentrations of extracellular 5-HT are deleterious to SSC development and suggest that transient 5-HTT expression in thalamocortical neurons is responsible for barrel patterns in neonatal rodents, and its permissive action is required for normal barrel pattern formation, presumably by maintaining extracellular 5-HT concentrations below a critical threshold. Because normal synaptic density in SSC layer IV of *Htt* KO mice was shown, it is more likely that 5-HT affects SSC cytoarchitecture by promoting dendritic growth toward the barrel hollows as well as by modulating cytokinetic movements of cortical granule cells, similar to concentration-dependent 5-HT modulation of cell migration described in other tissues. Since the gene-dose dependent reduction in 5-HTT availability in heterozygous KO mice, that leads to a modest delay in 5-HT uptake but distinctive irregularities in barrel and septum shape, is similar to those reported in humans carrying low activity regulatory variant of the *HTT*, it may be speculated that allelic variation in 5-HTT function also affects the human brain during development with due consequences for disease liability and therapeutic response (Schmitt & Lesch 2005).

These findings demonstrate that excessive amounts of extracellular 5-HT are detrimental to SSC development and suggest that transient 5-HTT expression and its permissive action is required for barrel pattern formation, presumably by maintaining extracellular 5-HT concentrations below a critical threshold. Two key players of serotonergic neurotransmission appear to mediate the deleterious effects of excess 5-HT: the 5-HTT and the 5-HT_{1B} receptor. Both molecules are expressed in primary sensory thalamic nuclei during the period when the segregation of thalamocortical projections occurs (Hansson et al 1998, Bennett-Clarke et al 1996). 5-HT is internalized via 5-HTT in thalamic neurons and is detectable in axon terminals (Lebrand et al 1996, Cases et al 1998). The presence of the VMAT2 within the same neurons allows internalized 5-HT to be stored in vesicles and used as a cotransmitter of glutamate. Lack of 5-HT degradation in *Maoa* KO mice as well as severe impairment of 5-HT clearance in mice with an inactivation of the *Htt* results in an accumulation of 5-HT and overstimulation of 5-HT receptors all along thalamic neurons (Cases et al 1998). Since 5-HT_{1B} receptors are known to inhibit the release of glutamate in the thalamocortical somatosensory pathway, excessive activation of 5-HT_{1B} receptors could prevent activity-dependent processes involved in the patterning of afferents and barrel structures. This hypothesis is supported by a recent study using a strategy of combined KO of *Maoa*, *Htt* and *Htr1b*. While only partial disruption of the patterning of somatosensory thalamocortical projections was observed in *Htt* KO, *Maoa*×*Htt* double KO (DKO) mice showed that 5-HT accumulation in the extracellular space causes total disruption of the patterning of these projections (Salichon et al 2001). Moreover, the removal of 5-HT_{1B} receptors in *Maoa* and *Htt* KO as well as in *Maoa*×*Htt* DKO mice allows a normal segregation of the somatosensory projections as well as retinal axons in the lateral geniculate nucleus (Upton et al 2002). These findings point to an essential role of the 5-HT_{1B} receptor in mediating the deleterious effects of excess 5-HT in the somatosensory system.

The effect of elevated extracellular 5-HT concentration on the modulation of programmed cell death during neural development was also investigated in early postnatal brains of *Htt* KO mice. *Htt* inactivation leads to a reduced number of apoptotic cells in striatum, thalamus, hypothalamus, cerebral cortex and hippocampus on postnatal day 1 (P1) with differences displaying an increasing frontocaudal gradient and regional specificity (Persico et al 2003). These findings underscore the role of 5-HT in the regulation of programmed cell death during brain development, and suggest that pharmacological enhancement of serotoninergic neurotransmission may minimize pathological apoptosis.

The evidence that changes 5-HT system homeostasis exerts long-term effects on cortical development and adult brain plasticity may be an important step forward in establishing the psychobiological groundwork for a neurodevelopmental hypothesis of negative emotionality, aggressiveness and violence (Lesch 2003a). Although there is converging evidence that serotonergic dysfunction contributes to anxiety-related behaviour, the precise mechanism that renders *Htt* KO mice more anxious and stress responsive as well as less aggressive remains to be elucidated.

Transcription factor Pet1

The comparatively small number of serotonergic neurons (~20 000 and ~300 000 in rodents and humans, respectively) are primarily located in the raphe nuclei, on the midline of the rhombencephalon, and in the reticular formation. Although these neurons are clustered in caudal and rostral divisions of the B1–B9 cell groups, the extensive collateralization of their terminals densely innervates all regions of the CNS. While serotonergic neurons are generated during early embryonic development (E10–E12), launch synthesis of 5-HT shortly after, and extend their axonal tracts to the forebrain and spinal cord, the maturational process shaping the networks is only completed during postnatal development. Despite the widespread importance of the central serotonergic neurotransmitter system, knowledge of the molecular mechanisms regulating the development of 5-HT neurons is still limited. The specification, differentiation, diversification, phenotype maintenance and survival of neurons comprising the raphe serotonergic system requires a considerable number of transcription factors, other morphogenetic regulators of gene expression, neurotrophins and growth factors as well as 5-HT itself to work in concert or in cascade (Lesch 2005).

Following definition of neuronal precursors' position several transcription factors turn out to be instrumental in establishing the serotonergic phenotype, reflected by the expression of gene representing the synthetic and metabolic machinery for 5-HT (e.g. tryptophan hydroxylase 2, MAOA), receptor-mediated signalling (e.g. 5-HT_{1A} receptor), or uptake-facilitated clearance (e.g. 5-HTT). Transcription factors that are expressed in postmitotic cells and that induce expression of these 5-HT markers encompass for example the ETS domain transcription factor Pet1 (Hendricks et al 1999, Ding et al 2003). Expression of Pet1 is restricted to the rostrocaudal extent of hindbrain raphe nuclei and is closely associated with developing serotonergic neurons in the raphe nuclei (Hendricks et al 1999, Pfaar et al 2002). Pet1 is therefore likely to be distinct from other factors because its expression pattern suggests that it performs a strictly serotonergic-specific function in the brain, Moreover, consensus Pet1 binding motifs are present in the transcriptional regulatory regions of both the human and murine genes for the 5-HT_{1A} receptor, 5-HTT, tryptophan hydroxylase 2 and aromatic L-amino acid decarboxylase (Aaad) whose expression profile is characteristic of the serotonergic neuron phenotype, i.e. 5-HT synthesis, release, uptake and metabolism.

In the rat dorsal and median raphe, 5-HT neurons begin to appear at approximately embryonic day (E11) and peaking at E13–E14. In these nuclei, it is thought that serotonergic neuron precursors begin to produce 5-HT near the time of their last cell division. The detection of Pet1 as early as E12.5 in the rostral cluster suggests that it is expressed in 5-HT neuron precursors during their terminal differentiation, consistent with its expression before the appearance of 5-HT. Taken together, these findings identify Pet1 as a critical regulator of serotonergic system specification.

While nearly all serotonergic neurons fail to differentiate in mice lacking Pet1, the remaining exhibit deficient expression of genes required for 5-HT synthesis, uptake and vesicular storage (Hendricks et al 2003). In target fields including cortex and hippocampus, 5-HT-specific fibres as well as 5-HT and 5-HIAA concentrations were also dramatically reduced in *Pet1* KO mice, whereas no major cytoarchitectural abnormalities in nuclear groups of several brain regions were detected. Interestingly, *Pet1* KO mice show evidence for increased anxiety-like behaviour in the Elevated Plus Maze test and enhanced aggressiveness in the Resident–Intruder test as a consequence of disrupted 5-HT system development. These findings further support the notion that Pet1 may represent the terminal differentiation factor which establishes the final identity of 5-HT neurons. Finally, the Pet1-dependent transcriptional program appears to couple 5-HT neuron differentiation during brain development to serotonergic modulation of behaviour related to anxiety and aggression in adulthood.

Monoamine oxidase A

Alterations in monoamine oxidase A (MAOA) activity have been implicated in a wide range of behavioural traits and disorders. MAOA is a mitochondrial enzyme that oxidizes 5-HT, norepinephrine as well as dopamine, and is expressed in a cell type-selective manner. *Maoa*-deficient mice were generated unintentionally by the replacement of exons 2 and 3 of the *Maoa* gene with an interferon transgene (Cases et al 1995). Mice with a targeted inactivation of the *Maoa* display elevated brain levels of 5-HT, norephinephrine and dopamine, increased reactivity to stress, hyperactive startle responses, violent motions during sleep and abnormal posture, and aggressive behaviour. Enhanced male aggressiveness was demonstrated by Resident–Intruder tests and by increased injury between male cage mates. The increased aggressiveness of the *Maoa* KO mice was indicated by the large percentage of mutant males that become wounded under standard group housing conditions and confirmed by enhanced offensive aggression by mutants in the Resident–Intruder test (Seif & De Maeyer 1999). *Maoa* KO mice also displayed increased copulatory behaviour of males with non-receptive female mice. Since these phenotypical alterations are restrained by 5-HT synthesis inhibition but not by catecholamine synthesis suppression, the observed behavioural abnormalities are likely to be specifically the result of attenuated 5-HT degradation. Since *Maoa* KO mice also show disrupted formation of sensory maps in the visual and somatosensory systems (cortical barrelfields) (Cases et al 1996, Salichon et al 2001), thus underscoring the role of 5-HT as a morphogenetic factor in brain development, it remains to be elucidated whether some of the behavioural abnormalities are influenced by structural abnormalities.

The aggressive phenotype of *Maoa* KO mice appears to complement the behavioural consequences of a mutation in the coding region of the human *MAOA*. This X-linked hemizygous chain termination mutation has been linked to mild mental retardation and occasional episodes of impulsive aggression, arson and hypersexual behaviour, such as attempted rape and exhibitionism, in affected males from a single large family (Brunner et al 1993). Affected males exhibit markedly disturbed monoamine metabolism and an absence of MAOA enzymatic activity in cultured fibroblasts. A non-conservative point mutation was found in all affected males and all carrier females; the mutation introduces a stop codon (base triplet, e.g. TAA, that serves as a signal for termination of transcription) at position 296. Although inhibition of MAOA in adults leads to antidepressant effects but not aggression-related behaviour, the deviate behaviour in MAOA-deficient men may be due to structural or compensatory changes resulting from altered monoamine metabolism during neurodevelopment. The fact that humans with an inactive MAOA gene also show increased impulsive aggression and sexual aggressiveness demonstrates the potential relevance of mutant mouse models to human behaviour, although the rarity of the human mutation indicates that other genetic and/or non-genetic influences that contribute to this forms of misconduct.

Although *MAOA* is a potential candidate for affective illness, none of several previously described gene variants are consistently associated with disease. A functional 30 basepair (bp) repeat polymorphism was identified in the promoter region of the human *MAOA* gene that differentially modulates gene transcription (Deckert et al 1999, Sabol et al 1998) as well as enzyme activity in fibroblasts. A corresponding functional repeat polymorphism was recently investigated with respect to its role in aggressive behaviour of rhesus monkeys (Newman et al 2005, also see Suomi 2005, this volume). Length variation of *MAOAs* upstream regulatory region confers vulnerability to antisocial behaviour in alcohol-dependent males (Samochowiec et al 1999), is linked to impulsivity, hostility and lifetime aggression history as well as brain serotonergic function in a community sample of men (Manuck et al 2000), and appears to be a risk factor for panic disorder and unipolar depression in female patients (Deckert et al 1999, Schulze et al 2000) but not for other psychiatric disorders (Syagailo et al 2001).

Gene versus environment interaction and emotionality

At the core of the gene versus environment debate, the relative influences of adverse experiences early in life on fear/anxiety and aggression is still a matter of intense debate. Investigations in rats have shown that maternal behaviour has long-lasting consequences on fear-related behaviour of the offspring. Maternal separation for several hours a day during the early postnatal period results in increased anxiety-like behaviours as well as increased stress responsivity in adult animals (Kalinichev et al 2002). Since the genetic basis of present-day temperamental and behavioural traits is already laid out in many mammalian species and may reflect selective forces among our remote ancestors, research efforts have recently been focused on non-human primates and humans.

Although studies in non-human primates have yielded useful models of aggressive behaviours, insights into the biological mechanisms that underlie these behaviours are barely beginning to emerge, and the concept of aggression as an antisocial instinct is being replaced by a framework that considers it a tool of competition and negotiation. Since the genetic basis of present-day personality and behavioural traits may reflect selective forces among our remote ancestors, research efforts have recently been focussed on rhesus macaques (also see Suomi 2005, this volume). In this non-human primate model environmental influences are probably less complex, can be more easily controlled for, and thus less likely to confound associations between temperament and genes. All forms of aggression in rhesus monkeys — major categories are defensive and offensive aggression — appear to be modulated by environmental factors, and marked disruptions to the mother–infant relationship likely confer increased risk for maladaptive aggression-related behaviour. In line with indirect evidence for influential effects of the quality of maternal care on life-long emotional behaviour depends on interactions between genes and the environment, allelic variation of 5-HTT function in rhesus macaques with histories of early-life stress was associated with altered central 5HT homeostasis, impulsive aggression, incompetent social behaviour, increased behavioural and endocrine responsivity to stress as well as greater risk to develop alcohol dependency (Bennett et al 2002, Barr et al 2004a,b).

Taken together, the findings provide evidence of an environment-dependent association between allelic variation of both 5-HTT and MAOA function, and illustrate the possibility that specific genetic factors play a role in 5-HT-mediated social competence in primates. Because rhesus monkeys exhibit temperamental and behavioural traits that parallel anxiety, depression and aggression-related personality dimensions associated in humans with low 5-HTT and MAOA function, it may be possible to search for evolutionary continuity in the genetic mechanism for individual differences as well as in gene×environment interactions underlying disease risk. Since the genetic basis of present-day temperamental and behavioural traits is already laid out in many mammalian species and may reflect selective forces among our remote ancestors, research efforts have recently been focused on humans (Caspi et al 2003) (also see Craig 2005, this volume).

Summary and future directions

Anxiety and aggression are complex social behaviours that arise out of multiple causes involving biological, psychological and social forces, and different forms of emotional behaviour may each result from different biopsychosocial pathways. The expression of anxiety and aggression must be carefully modulated to assure the success of individuals, small groups, and especially within large societies. Even though individual differences in emotionality and the behavioural consequences, such as violence, addiction and suicidality, are substantially heritable, they ultimately result from an interplay between genetic variations and environmental factors. Multiple neural networks, including the orbital frontal cortex, amygdala, anterior cingulate cortex and several other interconnected regions, whose formation, function and integration depend on the actions of several classical neurotransmitters, such as 5-HT, have been linked to anxiety and aggression. Several lines of evidence suggest that deficits in serotonergic modulation of this cortico-cingulo-amygdaloid circuit modify an individual's emotional response. Genetically influenced variation of 5-HT system function, in conjunction with other predisposing genetic factors and with inadequate adaptive responses to environmental stressors, is also likely to contribute to inappropriate anxiety- and aggression-related behaviour emerging from compromised brain development and from neuroadaptive processes. However, even pivotal regulatory proteins of neurotransmission, such as receptors, transporters, modifying enzymes and developmentally acting transcription factors will have only a modest impact, while noise from non-genetic mechanisms may seriously obstruct identification of relevant genes. Although current methods for the detection of gene×environment interaction in behavioural genetics are largely indirect, the most relevant consequence of gene identification for personality and behavioural traits related to aggression may be that it will provide the tools required to systematically clarify the effects of gene×environment interaction.

Based on the remarkable progress in technologies that allow the alteration or elimination of individual genes to create transgenic animal models, gene KO strategies are likely to increase our knowledge about which gene products are involved in anxiety- and aggression-related traits. However, because a missing gene might affect many developmental processes throughout ontogeny and compensatory mechanisms may be activated in KO mice, behavioural data from mice with targeted gene deletions should be interpreted with caution. It is becoming increasingly evident that many neurotransmitters and their receptors are expressed at early periods of neural development, and it is increasingly appreciated that they participate in the structural organization of the brain. An additional shortcoming of current KO experiments is the inability to provide region-specific control of the disruption. The ability to use native and exogenous promoters to control the expression of specifically targeted genes may allow region-specific and temporal control of protein expression. Systems that may prove useful include tetracycline or tamoxifen-responsive inducible promoters, and the loxP-Cre approach of inducibly deleting sections of DNA. The development of conditional KOs, in which a specific gene can be inactivated tissue-specifically any time during ontogeny, are therefore likely to avoid these imperfections associated with behavioural data from constitutive KOs.

References

Azmitia EC, Whitaker-Azmitia PM 1997 Development and adult plasticity of serotonergic neurons and their target cells. In: Baumgarten HG, Göthert M (eds) Serotonergic neurons and 5-HT receptors in the CNS. Springer, Berlin, Heidelberg, New York, p 1–39

Barnes NM, Sharp T 1999 A review of central 5-HT receptors and their function. Neuropharmacology 38:1083–1152

Barr CS, Newman TK, Lindell S et al 2004a Interaction between serotonin transporter gene variation and rearing condition in alcohol preference and consumption in female primates. Arch Gen Psychiatry 61:1146–1152

Barr CS, Newman TK, Schwandt M et al 2004b Sexual dichotomy of an interaction between early adversity and the serotonin transporter gene promoter variant in rhesus macaques. Proc Natl Acad Sci USA 101:12358–12363

Bengel D, Murphy DL, Andrews AM et al 1998 Altered brain serotonin homeostasis and locomotor insensitivity to 3, 4- methylenedioxymethamphetamine ("Ecstasy") in serotonin transporter-deficient mice. Mol Pharmacol 53:649–655

Bennett AJ, Lesch KP, Heils A et al 2002 Early experience and serotonin transporter gene variation interact to influence primate CNS function. Mol Psychiatry 7:118–122

Bennett-Clarke CA, Chiaia NL, Rhoades RW 1996 Thalamocortical afferents in rat transiently express high-affinity serotonin uptake sites. Brain Res 733:301–306

Blier P, de Montigny C 1998 Possible serotonergic mechanisms underlying the antidepressant and anti- obsessive-compulsive disorder responses. Biol Psychiatry 44:313–323

Brodkin ES 2005 Quantitative trait locus analysis of aggressive behaviours in mice. In: Molecular mechanisms of aggressive behaviour (Novartis Found Symp 268), p 57–77

Brunner D, Hen R 1997 Insights into the neurobiology of impulsive behavior from serotonin receptor knockout mice. Ann NY Acad Sci 836:81–105

Brunner HG, Nelen M, Breakefield XO, Ropers HH, van Oost BA 1993 Abnormal behavior associated with a point mutation in the structural gene for monoamine oxidase A. Science 262:578–580

Cases O, Vitalis T, Seif I et al 1996 Lack of barrels in the somatosensory cortex of monoamine oxidase A-deficient mice: Role of a serotonin excess during the critical period. Neuron 16:297–307

Cases O, Seif I, Grimsby J et al 1995 Aggressive behavior and altered amounts of brain serotonin and norepinephrine in mice lacking MAOA. Science 268:1763–1766

Cases O, Lebrand C, Giros B et al 1998 Plasma membrane transporters of serotonin, dopamine, and norepinephrine mediate serotonin accumulation in atypical locations in the developing brain of monoamine oxidase A knock-outs. J Neurosci 18:6914–6927

Caspi A, Sugden K, Moffitt TE et al 2003 Influence of life stress on depression: moderation by a polymorphism in the 5-HTT gene. Science 301:386–389

Cheetham SC, Crompton MR, Katona CL, Horton RW 1990 Brain 5-HT1 binding sites in depressed suicides. Psychopharmacology (Berl) 102:544–548

Chen JJ, Li Z, Pan H et al 2001 Maintenance of serotonin in the intestinal mucosa and ganglia of mice that lack the high-affinity serotonin transporter: abnormal intestinal motility and the expression of cation transporters. J Neurosci 21:6348–6361

Chiavegatto S, Dawson VL, Mamounas LA et al 2001 Brain serotonin dysfunction accounts for aggression in male mice lacking neuronal nitric oxide synthase. Proc Natl Acad Sci USA 98:1277–1281

Craig I 2005 The role of monoamine oxidase A, *MAOA*, in the aetiology of antisocial behaviour: the importance of gene environment interactions. In: Molecular mechanisms of aggressive behaviours (Novartis Found Symp 268). Wiley, Chichester, p 227–241

Crawley JN 1999 Behavioral phenotyping of transgenic and knockout mice: experimental design and evaluation of general health, sensory functions, motor abilities, and specific behavioral tests. Brain Res 835:18–26

Crawley JN, Paylor R 1997 A proposed test battery and constellations of specific behavioral paradigms to investigate the behavioral phenotypes of transgenic and knockout mice. Horm Behav 31:197–211

Davidson RJ, Putnam KM, Larson CL 2000 Dysfunction in the neural circuitry of emotion regulation-a possible prelude to violence. Science 289:591–594

De Vry J 1995 5-HT1A receptor agonists: recent developments and controversial issues. Psychopharmacology (Berl) 121:1–26

Deckert J, Catalano M, Syagailo YV et al 1999 Excess of high activity monoamine oxidase A gene promoter alleles in female patients with panic disorder. Hum Mol Genet 8:621–624

Di Pino G, Mössner R, Lesch KP, Lauder JM, Persico AM 2005 Serotonin roles in neurodevelopment: more than just neural transmission. Curr Neuropharmacol 2:403–417

Ding YQ, Marklund U, Yuan W et al 2003 Lmx1b is essential for the development of serotonergic neurons. Nat Neurosci 6:933–938

Drevets WC, Frank E, Price JC et al 1999 PET imaging of serotonin 1A receptor binding in depression. Biol Psychiatry 46:1375–1387

Fabre V, Beaufour C, Evrard A et al 2000 Altered expression and functions of serotonin 5-HT1A and 5-HT1B receptors in knock-out mice lacking the 5-HT transporter. Eur J Neurosci 12:2299–2310

Finn DA, Rutledge-Gorman MT, Crabbe JC 2003 Genetic animal models of anxiety. Neurogenetics 4:109–135

Flint J, Corley R, DeFries JC et al 1995 A simple genetic basis for a complex psychological trait in laboratory mice. Science 269:1432–1435

Flügge G 1995 Dynamics of central nervous 5-HT1A-receptors under psychosocial stress. J Neurosci 15:7132–7140

Gaspar P, Cases O, Maroteaux L 2003 The developmental role of serotonin: news from mouse molecular genetics. Nat Rev Neurosci 4:1002–1012

Gobbi G, Murphy DL, Lesch K, Blier P 2001 Modifications of the serotonergic system in mice lacking serotonin transporters: an in vivo electrophysiological study. J Pharmacol Exp Ther 296:987–995

Griebel G 1995 5-Hydroxytryptamine-interacting drugs in animal models of anxiety disorders: more than 30 years of research. Pharmacol Ther 65:319–395

Griebel G, Belzung C, Perrault G, Sanger DJ 2000 Differences in anxiety-related behaviours and in sensitivity to diazepam in inbred and outbred strains of mice. Psychopharmacology (Berl) 148:164–170

Groenink L, Pattij T, De Jongh R et al 2003 5-HT(1A) receptor knockout mice and mice overexpressing corticotropin-releasing hormone in models of anxiety. Eur J Pharmacol 463:185–197

Gross C, Zhuang X, Stark K et al 2002 Serotonin1A receptor acts during development to establish normal anxiety-like behaviour in the adult. Nature 416:396–400

Hansson SR, Mezey E, Hoffman BJ 1998 Serotonin transporter messenger RNA in the developing rat brain: early expression in serotonergic neurons and transient expression in non-serotonergic neurons. Neuroscience 83:1185–1201

Heisler LK, Chu HM, Brennan TJ et al 1998 Elevated anxiety and antidepressant-like responses in serotonin 5-HT1A receptor mutant mice [see comments]. Proc Natl Acad Sci USA 95:15049–15054

Hendricks T, Francis N, Fyodorov D, Deneris ES 1999 The ETS domain factor Pet-1 is an early and precise marker of central serotonin neurons and interacts with a conserved element in serotonergic genes. J Neurosci 19:10348–10356

Hendricks TJ, Fyodorov DV, Wegman LJ et al 2003 Pet-1 ETS gene plays a critical role in 5-HT neuron development and is required for normal anxiety-like and aggressive behavior. Neuron 37:233–247

Holmes A, Lit Q, Murphy DL, Gold E, Crawley JN 2003 Abnormal anxiety-related behavior in serotonin transporter null mutant mice: the influence of genetic background. Genes Brain Behav 2:365–380

Hoyer D, Martin G 1997 5-HT receptor classification and nomenclature: towards a harmonization with the human genome. Neuropharmacology 36:419–428

Kalinichev M, Easterling KW, Holtzman SG 2002 Early neonatal experience of Long-Evans rats results in long-lasting changes in reactivity to a novel environment and morphine-induced sensitization and tolerance. Neuropsychopharmacology 27:518–533

Korte SM, Meijer OC, de Kloet ER et al 1996 Enhanced 5-HT1A receptor expression in forebrain regions of aggressive house mice. Brain Res 736:338–343

Lauder JM 1993 Neurotransmitters as growth regulatory signals: role of receptors and second messengers. Trends Neurosci 16:233–240

Lebrand C, Cases O, Adelbrecht C et al 1996 Transient uptake and storage of serotonin in developing thalamic neurons. Neuron 17:823–835

Lemonde S, Turecki G, Bakish D et al 2003 Impaired repression at a 5-hydroxytryptamine 1A receptor gene polymorphism associated with major depression and suicide. J Neurosci 23:8788–8799

Lesch KP 2001 Mouse anxiety: the power of knockout. Pharmacogenomics J 1:187–192

Lesch KP 2003a Neuroticism and serotonin: a developmental genetic perspective. In: Plomin R, DeFries J, Craig I, McGuffin P (eds) Behavioral genetics in the postgenomic era. American Psychiatric Press, Washington, DC, p 389–423

Lesch KP 2003b The serotonergic dimension of aggression and violence. In: Mattson MP (ed) Neurobiology of aggression: understanding and preventing violence. Humana Press, Totowa, NJ, p 33–63

Lesch KP 2005 Genetic alterations of the murine serotonergic gene pathway: the neurodevelopmental basis of anxiety. In: Holsboer F, Ströhle A (eds) Anxiety and anxiolytic drugs (Handbook of Experimental Pharmacology). Springer, Berlin, Heidelberg, New York, p 71–112

Lesch KP, Mössner R 1999 5-HT1A receptor inactivation: anxiety or depression as a murine experience. Int J Neuropsychopharmacol 2:327–331

Lesch KP, Merschdorf U 2000 Impulsivity, aggression, and serotonin: a molecular psychobiological perspective. Behav Sci Law 18:581–604

Lesch KP, Murphy DL 2003 Molecular genetics of transporters for norepinephrine, dopamine, and serotonin in behavioral traits and complex diseases. In: Broeer S, Wagner CA (eds) Membrane transport diseases: molecular basis of inherited transport defects. Kluwer Academic/Plenum, New York, p 349–364

Lesch KP, Mayer S, Disselkamp-Tietze J et al 1990a 5-HT1A receptor responsivity in unipolar depression. Evaluation of ipsapirone-induced ACTH and cortisol secretion in patients and controls. Biol Psychiatry 28:620–628

Lesch KP, Disselkamp-Tietze J, Schmidtke A 1990b 5-HT1A receptor function in depression: effect of chronic amitriptyline treatment. J Neural Transm Gen Sect 80:157–161

Lesch KP, Hoh A, Schulte HM, Osterheider M, Muller T 1991 Long-term fluoxetine treatment decreases 5-HT1A receptor responsivity in obsessive-compulsive disorder. Psychopharmacology (Berl) 105:415–420

Lesch KP, Wiesmann M, Hoh A et al 1992 5-HT1A receptor-effector system responsivity in panic disorder. Psychopharmacology (Berl) 106:111–117

Lesch KP, Bengel D, Heils A et al 1996 Association of anxiety-related traits with a polymorphism in the serotonin transporter gene regulatory region. Science 274:1527–1531

Li Q, Wichems C, Heils A et al 1999 Reduction of 5-hydroxytryptamine (5-HT)(1A)-mediated temperature and neuroendocrine responses and 5-HT(1A) binding sites in 5-HT transporter knockout mice. J Pharmacol Exp Ther 291:999–1007

Li Q, Wichems C, Heils A, Lesch KP, Murphy DL 2000 Reduction in the density and expression, but not G-protein coupling, of serotonin receptors (5-HT1A) in 5-HT transporter knock-out mice: gender and brain region differences. J Neurosci 20:7888–7895

Lopez JF, Chalmers DT, Little KY, Watson SJ 1998 A.E. Bennett Research Award. Regulation of serotonin1A, glucocorticoid, and mineralocorticoid receptor in rat and human hippocampus: implications for the neurobiology of depression. Biol Psychiatry 43:547–573

Manuck SB, Flory JD, Ferrell RE, Mann JJ, Muldoon MF 2000 A regulatory polymorphism of the monoamine oxidase-A gene may be associated with variability in aggression, impulsivity, and central nervous system serotonergic responsivity. Psychiatry Res 95:9–23

Maxson SC 1996 Searching for candidate genes with effects on an agonistic behavior, offense, in mice. Behav Genet 26:471–476

Mayorga AJ, Dalvi A, Page ME et al 2001 Antidepressant-like behavioral effects in 5-hydroxytryptamine(1A) and 5-hydroxytryptamine(1B) receptor mutant mice. J Pharmacol Exp Ther 298:1101–1107

Murphy DL, Uhl GR, Holmes A et al 2003 Experimental gene interaction studies with SERT mutant mice as models for human polygenic and epistatic traits and disorders. Genes Brain Behav 2:350–364

Newman TK, Syagailo Y, Barr CS et al 2005 Monoamine oxidase A gene promoter polymorphism and infant rearing experience interact to influence aggression and injuries in rhesus monkeys. Biol Psychiatry 57:167–172

Olivier B, Mos J 1992 Rodent models of aggressive behavior and serotonergic drugs. Prog Neuropsychopharmacol Biol Psychiatry 16:847–870

Olivier B, Miczek KA 1999 Fear and anxiety: mechanisms, models and molecules. In: Dodman N, Shuster I (eds) Psychopharmacology of animal behavior disorders. Blackwell, London, p 105–121

Olivier B, Mos J, van Oorschot R, Hen R 1995 Serotonin receptors and animal models of aggressive behavior. Pharmacopsychiatry 28 Suppl 2:80-90

Olivier B, Soudijn W, van Wijngaarden I 1999 The 5-HT1A receptor and its ligands: structure and function. Prog Drug Res 52:103–165

Olivier B, Pattij T, Wood SJ et al 2001 The 5-HT(1A) receptor knockout mouse and anxiety. Behav Pharmacol 12:439–450

Parks CL, Robinson PS, Sibille E, Shenk T, Toth M 1998 Increased anxiety of mice lacking the serotonin1A receptor. Proc Natl Acad Sci USA 95:10734–10739

Parsons LH, Kerr TM, Tecott LH 2001 5-HT(1A) receptor mutant mice exhibit enhanced tonic, stress-induced and fluoxetine-induced serotonergic neurotransmission. J Neurochem 77: 607–617

Persico AM, Revay RS, Mössner R et al 2001 Barrel pattern formation in somatosensory cortical layer IV requires serotonin uptake by thalamocortical endings, while vesicular monoamine release is necessary for development of supragranular layers. J Neurosci 21:6862–6873

Persico AM, Baldi A, Dell'Acqua ML et al 2003 Reduced programmed cell death in brains of serotonin transporter knockout mice. Neuroreport 14:341–344

Pfaar H, von Holst A, Vogt Weisenhorn DM et al 2002 mPet-1, a mouse ETS-domain transcription factor, is expressed in central serotonergic neurons. Dev Genes Evol 212:43–46

Plomin R, Owen MJ, McGuffin P 1994 The genetic basis of complex human behaviors. Science 264:1733–1739

Ramboz S, Saudou F, Amara DA et al 1996 5-HT1B receptor knock out — behavioral consequences. Behav Brain Res 73:305–312

Ramboz S, Oosting R, Amara DA et al 1998 Serotonin receptor 1A knockout: an animal model of anxiety-related disorder. Proc Natl Acad Sci USA 95:14476–14481

Richer M, Hen R, Blier P 2002 Modification of serotonin neuron properties in mice lacking 5-HT1A receptors. Eur J Pharmacol 435:195–203

Rothe C, Gutknecht L, Freitag CM et al 2004 Association of a functional -1019C>G 5-HT1A receptor gene polymorphism with panic disorder with agoraphobia. Int J Neuropsychopharmacol 7:189–192

Sabol SZ, Hu S, Hamer D 1998 A functional polymorphism in the monoamine oxidase A gene promoter. Hum Genet 103:273–279

Salichon N, Gaspar P, Upton AL et al 2001 Excessive activation of serotonin (5-HT) 1B receptors disrupts the formation of sensory maps in monoamine oxidase A and 5-HT transporter knock-out mice. J Neurosci 21:884–896

Samochowiec J, Lesch KP, Rottmann M et al 1999 Association of a regulatory polymorphism in the promoter region of the monoamine oxidase A gene with antisocial alcoholism. Psychiatry Res 86:67–72

Santarelli L, Saxe M, Gross C et al 2003 Requirement of hippocampal neurogenesis for the behavioral effects of antidepressants. Science 301:805–809

Sargent PA, Kjaer KH, Bench CJ et al 2000 Brain serotonin1A receptor binding measured by positron emission tomography with [11C]WAY-100635: effects of depression and antidepressant treatment. Arch Gen Psychiatry 57:174–180

Saudou F, Amara DA, Dierich A et al 1994 Enhanced aggressive behavior in mice lacking 5-HT1B receptor. Science 265:1875–1878
Schmitt A, Lesch KP 2005 Serotonin transporter and treatment response: an update. Pharmacogenetics, in press
Schulze TG, Müller DJ, Krauss H et al 2000 Association between a functional polymorphism in the monoamine oxidase A gene promoter and major depressive disorder. Am J Med Genetics 96:801–803
Seif I, De Maeyer E 1999 Knockout Corner — Knockout mice for monoamine oxidase A. Int J Neuropsychopharmcol 2:241–243
Sibille E, Pavlides C, Benke D, Toth M 2000 Genetic inactivation of the Serotonin(1A) receptor in mice results in downregulation of major GABA(A) receptor alpha subunits, reduction of GABA(A) receptor binding, and benzodiazepine-resistant anxiety. J Neurosci 20:2758–2765
Sora I, Wichems C, Takahashi N et al 1998 Cocaine reward models: conditioned place preference can be established in dopamine- and in serotonin-transporter knockout mice. Proc Natl Acad Sci USA 95:7699–7704
Sora I, Hall FS, Andrews AM et al 2001 Molecular mechanisms of cocaine reward: combined dopamine and serotonin transporter knockouts eliminate cocaine place preference. Proc Natl Acad Sci USA 98:5300–5305
Stork O, Welzl H, Wotjak CT et al 1999 Anxiety and increased 5-HT1A receptor response in NCAM null mutant mice. J Neurobiol 40:343–355
Strobel A, Gutknecht L, Zheng Y et al 2003 Allelic variation of serotonin receptor 1A function is associated with anxiety- and depression-related traits. J Neural Transm 110:1445–1453
Suomi SJ 2005 Aggression and social behaviour in rhesus monkeys. In: Molecular mechanisms of aggressive behaviour (Novartis Found Symp). Wiley, Chichester, p 216–226
Syagailo YV, Stober G, Grassle M et al 2001 Association analysis of the functional monoamine oxidase A gene promoter polymorphism in psychiatric disorders. Am J Med Genet 105: 168–171
Tecott LH, Barondes SH 1996 Genes and aggressiveness. Behavioral genetics. Curr Biol 6: 238–240
Toth M 2003 5-HT(1A) receptor knockout mouse as a genetic model of anxiety. Eur J Pharmacol 463:177–184
Upton AL, Ravary A, Salichon N et al 2002 Lack of 5-HT(1B) receptor and of serotonin transporter have different effects on the segregation of retinal axons in the lateral geniculate nucleus compared to the superior colliculus. Neuroscience 111:597–610
Vitiello B, Stoff DM 1997 Subtypes of aggression and their relevance to child psychiatry. J Am Acad Child Adolesc Psychiatry 36:307–315
Weiss SM, Lightowler S, Stanhope KJ, Kennett GA, Dourish CT 2000 Measurement of anxiety in transgenic mice. Rev Neurosci 11:59–74
Wisor JP, Wurts SW, Hall FS et al 2003 Altered rapid eye movement sleep timing in serotonin transporter knockout mice. Neuroreport 14:233–238
Wissink S, Meijer O, Pearce D, van Der Burg B, van Der Saag PT 2000 Regulation of the rat serotonin-1A receptor gene by corticosteroids. J Biol Chem 275:1321–1326
Zhuang X, Gross C, Santarelli L et al 1999 Altered emotional states in knockout mice lacking 5-HT1A or 5-HT1B receptors. Neuropsychopharmacology 21:52S–60S

DISCUSSION

Olivier: In overviewing the receptors and transporters, you had the most specificity in the receptors. I'm assuming that your data from humans on the 5-HT_{1A} receptor, for example, are coming from blood cells.

Lesch: This is not our work, it is by Paul Albert's group from Ottawa, Canada. They have studied the promoter primarily in *in vitro* systems, using reporter gene assays and binding studies of transcription factors to this particular SNP.

Olivier: But you know that the 5-HT_{1A} and $_{1B}$ receptors are both presynaptic and postsynaptic. Also, there are not good data to support a differential function. It could be that both are regulated in a different way. I believe that particularly aggression is more regulated on the 5-HT_{1B} postsynaptic site than on the presynaptic site. The 5-HT_{1B} receptor is the one to study, because if you induce the serotonergic system by this transcription factor you see enhanced aggression. In your 5-HT transporter knockouts you see an enormously enhanced level of serotonin, which leads to reduced aggression. I think 5-HT_{1B} receptor is the important receptor for aggression.

Lesch: You have made an important point: we need to look at where the receptor is expressed. Of course, there are considerable variations in the brain regions with regard to the postsynaptic 5-HT_{1B} versus the 5-HT_{1A} receptor expression. This might explain many of these discrepant findings. The $_{1B}$ is probably more related to the aggressive behaviour and $_{1A}$ more to anxiety-related behaviour. There have been some interesting rescue studies showing that the postsynaptic expression of $_{1A}$, depending on the stage of brain development, is important for the modulation of anxiety in later life. Similar studies could be done for $_{1B}$ and would tease out its importance for aggression-related behaviour.

Keverne: As well as the pre- and postsynaptic differences, what about the regional differences? Is there anywhere in the brain where $_{1B}$ is specifically expressed and not $_{1A}$ for example?

Lesch: I think 5-HT_{1B} is primarily in striatum. It is more related to the dopamine system than any other 5-HT receptor.

Keverne: This fits in part with addiction. In humans, addiction, alcohol intake, heroin withdrawal and cocaine are all associated with impulsive aggression.

Suomi: Do you want to say something about the interaction between MAOA and the serotonin transporter?

Lesch: We haven't seen any interaction with regard to these traits. It is very difficult in human genetics looking at candidate genes to tease out any interactions. Most of the time it is a problem of statistical power. You need huge sample sizes, well beyond 1000 individuals. Very few investigators have patient samples of this size. So far with regard to personality traits we have not seen any interaction with MAOA and serotonin transporters. We have seen interactions with candidate genes from other systems, such as the dopamine transporter and brain-derived neurotrophic factor (BDNF).

Interestingly, we have not seen anything with genes of the serotonergic pathway. This is puzzling, because we were very much expecting to see this. In humans we hoped that if we combined two major players of serotonergic function and looked at them in a regression model that we would find interactions, but so far we haven't.

Craig: Part of the problem with that is that there are sex-specific effects. MAOA is specific for males and my gut feeling is that the effects of a low activity serotonin transporter is mainly in the females.

Brodkin: In mouse models, there are several different tests of anxiety-related behaviour, such as the elevated plus maze, the light–dark box and open field behaviour. Behaviour in these different anxiety-related tests is not always perfectly correlated, and patterns of inbred strain differences are not exactly the same in different tests of anxiety. There are also different tests of aggressive behaviour. In these knockouts when you say anxiety is increased or decreased, do you see consistent patterns across multiple tests?

Lesch: The 5-HT transporter knockout mice have been bred back into three different strains, CD1, 129/SV and C57BL/6J. The anxiety-like response was very similar in all three backgrounds, as it was seen in 5-HT_{1A} knockout mice previously. With regard to low aggression, it has only been studied in the C57BL/6J background so far, but the effect was quite robust. I guess this would also hold up in other backgrounds. So the genetic background is not an issue. There may be other confounding variables involved.

Brodkin: I didn't mean the genetic background. I meant different behavioural tests of anxiety.

Lesch: For the 5-HT transporter knockouts we did a quite comprehensive set of tests. Most notably we saw a gender-specific effect. Females always showed the strongest effect, and this was nicely in line with the human findings. Aggressiveness associated with 5-HT transporter function is gender specific and much more prominent in females.

Brembs: Given the strong developmental effects you see in humans, would you exclude that the difference you see between the mouse model system and humans is caused entirely by developmental processes? If you can't exclude this possibility, how does that make you feel in terms of the reliability of your research? How cautious does it make you in your interpretation of these results?

Lesch: If we knew it, we wouldn't call it research! I strongly suspect that the neurodevelopmental aspect is extremely important. But keep in mind that variation in serotonergic gene function is already there from the time of conception. Allelic variation of 5-HT transporter function during brain development is setting the stage for certain behaviours in later life. The constitutive 5-HT transporter knockout mouse might be the better model than the conditional knockout, where a gene is knocked out later in development. In some aspects it might be good to knock out the gene from the earliest point of development and model this. We do not have a human knockout for the 5-HT transporter, except for one family in which males are missing the MAOA gene. It might be worth looking at the heterozygous 5-HT transporter knockout mice more closely, which have a 50% decrease of 5-HT transporter expression. This very much reflects what we see with the low activity 5-HT transporter promoter genotype in humans. We have studied this quite extensively and have seen a subtle but consistent phenotype in heterozygotes in several paradigms of 5-HT function.

Manuck: Don gave us some examples yesterday of a gene that may vary appreciably in its phenotypic effects and speculated on the nature of the aggressive behaviour with which the gene might be associated. Perhaps this is relevant to discussion of regulatory variation in the serotonin transporter gene, as my reading suggests this literature is quite mixed. In contrast to studies favouring an association of aggression with the 5-HTTLPR short allele, homozygosity for the long allele has been associated with reported violence in suicidal adolescents (Zalsman et al 2001), with maternal reports of aggressive behaviour in offspring of alcoholic fathers (Twitchell et al 2001), and with adolescent aggressive and antisocial behaviour among adoptees of biological parents characterized as antisocial (Cadoret et al 2003). Other data suggest, too, that the long allele may predominate in antisocial alcoholics (Parsian & Cloninger 2001) and that among patients having Alzheimer's disease, the long allele predicts aggressive behaviour observed during the course of dementia (Sukonick 2001, Sweet et al 2001). Although I realize these studies involve diverse clinical populations, many with relatively small samples, do you believe the discrepant findings emerging in this literature are potentially reconcilable?

Lesch: The problem is always the instrument we use to measure human behaviour and personality. But the more recent studies focusing on endophenotypes such as amygdala responsivity, or looking at gene–environment interactions are now showing much stronger 5-HT transporter gene effects. If measures other than personality inventories you get much more robust effects and see a stronger impact of allelic variation. This is the direction the research should take.

Manuck: This would explain a mixed literature with many null effects. What I find a bit perplexing is a competing literature that finds positive associations for homozygosity in the 5-HTTLPR long allele.

Lesch: We just look at one strong regulatory mechanism of gene expression. There may be many others. There may be SNPs in the regulatory regions, or variations in introns of the genes. This is just a crude way of looking at allelic variation of expression and function. There might even be two promoters in the serotonin transporter gene, one regulating the expression in the gut versus the other one regulating expression in the brain, and future studies have to address the considerable complexity of gene regulation. Some of the discrepancies in the literature might be due to these complicated regulatory events. Our approach is just the first step.

Craig: I would like to endorse the remarks you were making about the promoter variance, and the problems in the literature with non-replication, or effects in the opposite direction. We are tending to assume that this variation in the promoter is simple, but as you said we know there are SNPs and the repeats aren't all the same. It is possible that one of the long variants has low activity. It could really confuse the issue.

Lesch: As a matter of fact, as we get more details on the regulation, the associations become stronger.

Lovell-Badge: We have to be careful with any regulatory mutation. In an *in vitro* test the readout is up or down, but in developmental time all sorts of things could be happening. It could be a normal level of expression until you get to a fully differentiated state, and then you see a reduction. If you are going to model it, you should try to recreate the same sort of mutation in your mice.

Lesch: You could end up with a mouse with no phenotype after investing years of work. This is a problem. Theoretically one should be able to model this 5-HT transporter gene promoter variation in mice, but again, it would be very labour-intensive. We now investigate a functional variation in the coding region in mice, which apparently increases uptake function and appears to be associated with a complex serotonergic phenotype. This is in the gene's coding region, and it is much more straightforward than modelling something in the regulatory region, because there is no sequence homology in the promoter apart from in the core promoter approximately 100 bases upstream of the transcription start site. When you study length variation of transcriptional control regions in different species, you never know whether the same transcription factors are modulating expression.

Manuck: I appreciate your attempt to develop a model that encompasses both the externalizing phenotypes of aggression and more internalized, anxiety-related phenotypes within a common framework implicating gene-specific variation. There is another, perhaps complementary way of thinking about the affective correlates of central serotonergic function, though perhaps not yet in the context of genetic variation. Zald & Depue (2001) reported an interesting study in which they assessed the prolactin response to fenfluramine among healthy individuals who were subsequently asked to record their mood states repeatedly during the course of daily life. On analysis of subjects' ambulatory mood reporting, Zald & Depue (2001) found that a blunted central serotonergic response predicted aggregated experiences of both positive *and* negative affect. Now several theories postulate independent axes of positive and negative emotionality as primary dimensions of human personality, with these two dimensions serving as substrates for different motivational systems (e.g. approach, appetitive, motivation and threat avoidance) (e.g. Watson 2000). Most taxonomies of personality and temperament also posit a third dimension of individual differences that, unlike the first two, is not believed to underlie a particular motivational system. Instead, this third dimension reflects variability in the central regulation of affect and impulse, and significantly, is thought to covary inversely with brain serotonergic activity. On this view, diminished serotonergic function may potentiate aggression, but would do so indirectly (perhaps by enhancing sensitivity to cues of threat or punishment) rather than by directly triggering aggressive acts. And as shown by Zald & Depue (2001), low serotonergic function might equally promote positive affect and behaviours motivated by positive affect, as by increasing sensitivity to reward-related environmental cues. If true — and not all evidence supports such speculation (e.g. Flory et al 2004) — serotonin may be more the arc and swing of emotion's pendulum than the pendulum itself.

Lesch: This once again points out the difficulties of these assessment systems. Up till now the NEO personality inventory (encompassing the 'Big Five' personality dimensions) has been the most consistent in the hands of other investigators, but it is not the best we would like to have. I would like to reiterate that we need other measures of endophenotypes. Perhaps electrophysiological measures would be useful. In our lab we have shown robust effects in small sample sizes with an $n=30$. fMRI has been used, as have paradigms of cognitive measures that show strong effects. Personality assessment is the first approximation, but not the way to go in the future.

Manuck: Is serotonin primarily associated with a particular quality or valence of affect, or with something else, something that modulates the expression of affect (whatever its valence)?

Lesch: As you put the question, it is not clear.

Pfaff: Whenever I see both anxiety and aggression going up, I am thinking of the underlying dimension of generalized arousal!

References

Cadoret RJ, Langbehn D, Caspers K et al 2003 Associations of the serotonin transporter promoter polymorphism with aggressivity, attention deficit, and conduct disorder in an adoptee population. Comprehensive Psychiatry 44:88–101

Flory JD, Manuck SB, Matthews KA, Muldoon MF 2004 Serotonergic function in the central nervous system is associated with daily ratings of positive mood. Psychiatry Res 129:11–19

Parsian A, Cloninger CR 2001 Serotonergic pathway genes and subtypes of alcoholism: association studies. Psychiatry Genet 11:89–94

Sukonick DL, Pollock BG, Sweet RA et al 2001 The 5-HTTLPR*S/*L polymorphism and aggressive behaviour in Alzheimer disease. Arch Neurol 58:1425–1428

Sweet RA, Pollock BG, Sukonick DL et al 2001 The 5-HTTLPR polymorphism confers liability to a combined phenotype of psychotic and aggressive behaviour in Alzheimer disease. International Psychogeriatrics 13:401–409

Twitchel GR, Hanna GL, Cook EH, Stoltenberg SF, Fitzgerald HE, Zucker RA 2001 Serotonin transporter promoter polymorphism genotype is associated with behavioural disinhibition and negative affect in children of alcoholics. Alcohol Clin Exp Res 25:953–959

Watson D 2000 Mood and temperament. Guilford Press, New York

Zald DH, Depue RA 2001 Serotonergic functioning correlates with positive and negative affect in psychiatrically healthy males. Personality Individ Diff 30:71–86

Zalsman G, Frisch A, Bromberg M et al 2001 Family-based association studies of serotonin transporter promoter in suicidal adolescents: No association with suicidality but possible role in violence traits. Am J Medl Genet (Neuropsychiatric Genet) 105:239–245

Effects of nitric oxide on the HPA axis and aggression

Randy J. Nelson

Departments of Psychology and Neuroscience, The Ohio State University, Columbus, OH 43210, USA

Abstract. We previously documented that male mice lacking the gene encoding the neuronal isoform of nitric oxide synthase (nNOS$^{-/-}$) are more aggressive than wild-type (WT) mice in all standard testing paradigms. Testosterone is necessary, but not sufficient, to evoke the persistent aggression in these mutants. Deletion of the nNOS gene not only eliminates nNOS protein, but in common with many gene deletions, affects several 'down-stream' processes. For example, serotonin (5-HT) metabolism is altered in male nNOS$^{-/-}$ mice. Baseline corticosterone, but not ACTH, concentrations are also altered in the nNOS$^{-/-}$ mice. Despite elevated corticosterone concentrations, nNOS knockout mice are less 'anxious' or 'fearful' than WT mice, which may contribute to their aggressiveness. For example, male nNOS$^{-/-}$ mice spend more time in the open field than WT mice. Furthermore, nNOS knockout mice also show increased sensitivity to painful stimuli, which may also prolong aggressive interactions. Aggressive behaviour is not a unitary process, but is the result of complex interactions among several physiological, motivational, and behavioural systems, with contributions from the social and physical environment. The multiple, and often unanticipated, effects of targeted gene disruption on aggressive behaviour are discussed.

2005 Molecular mechanisms influencing aggressive behaviours. Wiley, Chichester (Novartis Foundation Symposium 268) p 147–166

Aggressive behaviour is highly complex, and serves several adaptive functions. Aggression is also observed in many pathological or nonadaptive settings as well. Aggression has been defined as overt behaviour with the intention of inflicting physical damage upon another individual (Moyer 1971). The possibility for aggressive behaviour exists whenever the interests of two or more individuals conflict (Svare 1983). Conflicts usually arise over limited resources including territories, food and mates. Indeed, the ubiquitous resident–intruder aggression test models rodent territorial aggression. In nature, the social interaction decides which animal gains access to the contested resource. In many cases, a submissive posture or gesture on the part of one animal avoids the necessity for actual combat over a resource. Animals may also participate in psychological intimidation by engaging in threat displays or ritualized combat in which dominance is determined, but no physical damage is inflicted (Moyer 1971).

The likelihood of a single aggression brain circuit underlying all forms of aggressive behaviour seems remote. Nonetheless, a major goal of aggression research from a biomedical or veterinarian medical perspective is to develop pharmacological interventions suitable to control aggression in violent individuals. The results of this endeavour have been mixed. The best example of successful pharmacological control of aggression was realized over a half century ago with the development and use of neuroleptics (Miczek & Fish 2005). These pharmaceuticals dramatically changed how clinicians manage violence in people suffering from one or more disorders including psychotic, depressive, intellectual, or organic brain disorders, as well as managing violence in people abusing alcohol or amphetamine (Citrome & Volavka 1997a,b). Indeed, the effectiveness of chlorpromazine and haloperidol in reducing violent behaviour continues to set the standard in the evaluation of novel compounds for this purpose (Connor et al 2003, Miczek & Fish 2005). Although the first-generation neuroleptics worked primarily by sedating the patients and suppressing most behaviours, more recently developed drugs produce fewer negative side-effects such as tardive dyskinesia during chronic treatment (Swann 2003).

One simplified neurochemical model of aggressive behaviour suggests that serotonin inhibits, whereas dopamine permits or enables, aggressive behaviour. Recently, γ-aminobutyric acid (GABA) has been included in this model. Pharmacological interventions have been less successful when based upon this simple model. For example, although increasing serotonin via specific serotonin reuptake inhibitors (SSRIs) decreases aggressive behaviours in the majority of laboratory or clinical settings (Walsh & Dinan 2001, New et al 2004), a significant minority of individuals treated with SSRIs *increase* aggression (Spigset 1999). Similarly, although benzodiazepine, which enhances ligand binding to $GABA_A$ receptors, reduces aggression in most individuals, a significant minority display a paradoxical elevation in violence after benzodiazepine treatment (e.g. Azcarate 1975). Such individual differences are critical for our understanding and targeting of any successful pharmacological treatment for violence.

Clearly, a very specific pharmacological approach will be necessary to regulate violence in veterinarian or human medicine. Our laboratory has been studying the role of nitric oxide (NO), a signalling molecule in the nervous system, in the regulation of aggression. In this chapter, the behavioural effects of genetically or pharmacologically depleted NO are reviewed. In addition to the dramatic aggressive behavioural phenotypic effects, we have recently discovered several subtle effects of NO that interact to affect aggressive behaviour. For example, mice missing NO in neurons are less 'anxious' or 'fearful' than wild-type mice, traits that may contribute to their aggressiveness. Mice lacking NO from neurons also display increased sensitivity to painful stimuli, which may also affect aggressive interactions. Aggressive behaviour is not a unitary process, but is the result of complex interactions among several physiological, motivational and behavioural systems, with contributions from the social and physical environment. Taken together, any pharmacological tools developed to reduce aggression must likely affect several molecular signalling pathways simultaneously to be effective.

Nitric oxide

Nitric oxide (NO) was initially identified as an endogenous regulator of blood vessel tone (Ignarro 1990, Moncada & Higgs 1993). NO also mediates the bactericidal and tumoricidal actions of macrophages (Nathan 1992, Nathan & Xie 1994), and acts as a neuronal messenger in the central and peripheral nervous systems (Dawson & Snyder 1994, Dawson & Dawson 1996). NO is an endogenous gas that has biochemical properties of a free radical. NO is very labile, with a half-life of <5 s; consequently, many studies have manipulated NO indirectly by affecting its synthetic enzyme, nitric oxide synthase (NOS), that transforms arginine into citrulline and NO. Three distinct isoforms of NOS have been discovered in:

- the endothelial tissue of blood vessels (eNOS or NOS-3),
- macrophages as an inducible form (iNOS or NOS-2), and
- neurons (nNOS or NOS-1).

Suppression of NO formation by either elimination of arginine or by *N*-methyl-arginine, a potent NOS inhibitor, affects all three isoforms of NOS. Neurons containing nNOS are discretely localized throughout the brain. High densities of nNOS-positive cells are localized in the limbic system, which regulates emotional behaviour and aggression, particularly the lateral septal nuclei, posterior hypothalamus, entorhinal cortex and amygdala (Albert & Walsh 1984, Vincent & Kimura 1992).

During the establishment of a breeding colony of mice lacking the nNOS gene (nNOS$^{-/-}$), we conducted a series of behavioural studies (Nelson et al 1995). In both studies of aggression and mating behaviour, male nNOS$^{-/-}$ mice exhibited a dramatic loss of behavioural inhibition reflected by persistent fighting and mounting behaviour despite obvious [to the human observers] signals of surrender or disinterest, respectively, by their test partners. Nulliparous female nNOS$^{-/-}$ mice displayed neither elevated aggressiveness, nor inappropriate mating behaviours. Prolonged aggressiveness and mating behaviour among males are often associated with elevated blood concentrations of testosterone

(Simon 2002); however, no differences in testosterone concentrations were detected between wild-type (WT) and nNOS$^{-/-}$ males (Nelson et al 1995). However, castration of nNOS$^{-/-}$ mice results in a marked reduction in aggression and testosterone replacement therapy restores aggression to pre-castration levels (Kriegsfeld et al 1997). These results suggest that testosterone is necessary, but not sufficient, to elevate aggression in nNOS$^{-/-}$ mice. No sensory or motor deficiencies were observed that could account for the elevated aggression or mounting behaviour among the nNOS$^{-/-}$ mice.

Male reproductive and aggressive behaviours are both generally modulated, if not regulated, by androgens, presumably because defence of resources and competition are critical for reproductive success (Simon 2002). However, non-gonadal mechanisms may have evolved to regulate aggression in animals living in habitats that require competition outside of the breeding season (e.g. Soma & Wingfield 2001). Siberian hamsters (*Phodopus sungorus*) display an 'atypical' seasonal pattern of aggressive behaviour. Males that reduce gonadal size and function after exposure to short, winter-like days have low circulating testosterone concentrations, but display elevated aggression as compared to animals with large functional gonads and relatively high testosterone values (Jasnow et al 2000).

In a recent study, nNOS expression in the brains of long- and short-day Siberian hamsters (*Phodopus sungorus*) was examined after assessment of aggressive behaviour (Wen et al 2004). The reproductive response to short days varies in this species: short-day responsive hamsters inhibit reproductive function and have undetectable testosterone concentrations, whereas short-day non-responsive hamsters display fully functional gonads and long-day testosterone blood concentrations. Regardless of gonadal response to short days, all hamsters housed in short photoperiods were more aggressive than long-day animals. These results indicate that the short-day mediated aggression is not mediated by testosterone, because testis size and testosterone concentrations of the short day non-responders were similar to long-day animals. Short-day Siberian hamsters, again regardless of reproductive response, also displayed significantly fewer nNOS-immunoreactive cells in several amygdala regions compared to long-day animals. Together, these results suggest seasonal aggression in male Siberian hamsters is regulated by photoperiod, probably independently from gonadal steroid hormones, and may be regulated by nNOS.

Female mice do not typically show much territorial aggression, and aggression in nulliparous female nNOS$^{-/-}$ mice was equivalent to WT animals. However, female mice are more likely to engage in aggression when guarding their nest. In contrast to our predictions, female nNOS$^{-/-}$ mice were much less likely to display maternal aggression than WT females (Gammie & Nelson 1999, Gammie et al 2000). Again, there were no dramatic sensorimotor deficits among the mutant mice to account for the changes in aggressive behaviour. Taken together, these results suggest that neuronal NO has important, but opposite, effects in the mediation of aggression in male and female mice. Although there are no sex differences in NOS activity in the cortex, cerebellum, amygdala, or hypothalamus, androgens generally inhibit and estrogens generally increase NOS activity in the brain (Singh et al 2000, Weiner et al 1994).

Cautionary aside regarding knockout mice

A conceptual problem with behavioural studies of mice with targeted genetic deletions, that is shared with all ablation studies, is that behavioural tests reveal the effects of the *missing* gene (and gene product), not the effects of the gene directly (Nelson 1997). This conceptual shortcoming can be overcome in the same way that it is overcome in other types of ablation studies, by collecting convergent evidence from a variety of methods. For example, if similar behavioural deficits are obtained after pharmacological, lesion and genetic manipulation of the same factor, then it is reasonable to conclude that the missing factor is involved in the behaviour, especially if the behavioural deficit is ameliorated when the missing factor is restored. One significant advantage to using knockout animals in behaviour research is that the effects of the gene product can be abolished without the side-effects of drugs (Crawley 2000). This is particularly important in manipulation of NO because many drugs that non-specifically block synthesis of NO affect nitrogen-dependent physiological processes, and have pronounced effects on blood pressure (Dawson & Snyder 1994, Dawson & Dawson 1996). Such physiological effects can obviously confound behavioural processes mediated by NO. It is also important to validate behavioural findings on mutant mice on 'real' animals.

In order to investigate whether the increased aggressive behaviour of the mutants was due to the missing gene during the development of the brain with subsequent activation of compensatory mechanisms (Nelson 1997), WT male mice were treated with 7-nitroindazole (7-NI) (50 mg/kg ip), a relatively specific drug that blocks nNOS activity *in vivo* (Demas et al 1997). Indeed, a dramatic reduction of NOS activity in brains of 7-NI-treated animals was revealed by immunocytochemical staining for citrulline, an indirect marker for the NO synthesis (Demas et al 1997). Male mice treated with 7-NI exhibited substantially increased aggressive behaviour in two different tests as compared to control animals, with no alteration in other locomotor activities, implying an specific effect in aggression and ruling out the contribution of strain differences in the knockout mice in the aggressive behavioural phenotype (cf. Le Roy et al 2000). These pharmacological data extend the behavioural results obtained in $nNOS^{-/-}$ mice and confirm a role of NO in aggression.

eNOS and aggression

We hypothesized that NO from the endothelial tissue could contribute to aggressive behaviour. Mice with targeted disruption of the gene encoding the endothelial isoform of nitric oxide synthase (eNOS$^{-/-}$) were examined (Huang et al 1995, Demas et al 1999). Because NO was originally identified as endothelium-derived relaxing factor (Ignarro 1990), and eNOS is localized in the endothelial lining of vascular smooth muscle, affected blood pressure was the first phenotype investigated in these mutant mice (Huang et al 1995). eNOS$^{-/-}$ mice exhibit ~35% increase in basal blood pressure relative to WT mice (Huang et al 1995).

Anecdotal observations indicated that eNOS knockout mice were very docile. Animals were tested using two behavioural paradigms. First, animals were tested using the resident-intruder paradigm; eNOS$^{-/-}$ mice displayed virtually no aggressive encounters and a dramatic decreased duration of agonistic encounters relative to WT mice when a WT intruder was placed into their home cage. Likewise, when tested in a neutral arena with a WT stimulus male, eNOS$^{-/-}$ mice displayed many fewer attacks and a greatly increased latency to attack the stimulus male relative to WT mice (Demas et al 1999). Pharmacological normalization of blood pressure did not affect the absence of aggression in eNOS$^{-/-}$ mice. These data, in combination with the nNOS$^{-/-}$ data, suggest that the two isoforms of NOS may normally act to increase (eNOS$^{-/-}$) and decrease (nNOS$^{-/-}$) aggressive behaviour *in vivo*. Thus, WT mice with normal concentrations of both isoforms of NOS display only moderate levels of aggression. Throughout the proposed studies, we intend to compare the eNOS$^{-/-}$ mice with the nNOS$^{-/-}$ and WT animals.

Serotonin and aggression

Both pharmacological and clinical approaches have identified serotonin (5-HT) as an important neurotransmitter system involved in aggression and impulsivity. Elevated aggression in humans is correlated with low cerebrospinal fluid concentrations of 5-hydroxyindoleacetic acid (5-HIAA) (5-HT metabolite) and a blunted prolactin response to a 5-HT agonist (fenfluramine challenge) (reviewed in Manuck et al 1998, Nelson & Chiavegatto 2001, Miczek & De Almeida 2001). Aggressive laboratory animals display reduced brain 5-HT turnover (reviewed in Nelson & Chiavegatto 2001, Miczek & De Almeida 2001). The inverse relationship between 5-HT system activity and aggressive behaviour may be causally linked as suggested by pharmacological manipulation of brain 5-HT concentrations. Drugs that inhibit 5-HT synthesis, tryptophan-free diets, or lesions of 5-HT neurons decrease 5-HT levels and elevate aggression in laboratory animals (reviewed in Chiavegatto et al 2001). In contrast, decreased aggression occurs after treatment with 5-HT precursors, 5-HT reuptake inhibitors and 5-HT releasing agents, in addition to 5-HT1A and 5-HT1B agonists (reviewed in Olivier et al 1995, Chiavegatto et al 2001, Miczek & De Almeida 2001, De Almeida et al 2001). Gene targeting strategies in mice that either directly or indirectly affect the functional integrity of the 5-HT system have generally supported the hypothesis that 5-HT modulates aggression (reviewed by Nelson & Chiavegatto 2001, Miczek et al 2001).

Serotonin neurotransmission was hypothesized to be disrupted in the aggressive nNOS null mice because of the inverse relation of 5-HT system activity and aggression. Serotonin metabolism, analysed by the ratio of the metabolite 5-HIAA and the 5-HT levels by HPLC, was significantly reduced in several brain regions including the cortex, hypothalamus, midbrain, and cerebellum of $nNOS^{-/-}$ in comparison to WT mice (Chiavegatto et al 2001). The alterations in 5-HT turnover were due to increased concentrations of 5-HT with no changes in its metabolite in most brain regions studied. The disturbed neurochemical profile appears specific to the 5-HT system, because norepinephrine, dopamine and metabolites were generally unaffected. Taken together, these data suggest that it is unlikely that alterations in monoamine oxidase account for the 5-HT abnormalities in the nNOS knockout mice (Chiavegatto & Nelson 2003).

HPA axis, anxiety and aggression

Anxiety levels and aggressive behaviour are indirectly related. There is evidence that elevated aggression is related to low anxiety (reviewed in Olivier et al 1995, Miczek et al 2003). For example, genetically selected aggressive mice display fewer anxiety-like behaviours on standard tests than mice selected to be nonaggressive (Nyberg et al 2003). There is also evidence that high anxiety provokes elevated aggression (reviewed in Blanchard et al 2001). For example, unpredictable chronic mild stress increased aggression both in a resident–intruder test and between cage-mates (Mineur et al 2003). Behavioural assessment of adenosine A1 receptor knockout mice ($A1R^{-/-}$) revealed elevated anxiety-like behaviours and increased aggressiveness in the resident–intruder test (Givalois-Llort et al 2002). The relationships between anxiety and aggression appear to be primarily modulated at the level of the GABA receptor system (Miczek et al 2003).

Considerable evidence in rodents suggests the HPA regulates both stress responses and aggression (Haller & Kruk 2003). Both increased and decreased activity of the HPA axis can affect aggressive behaviour. Chronic activation of the HPA axis and the subsequent release of glucocorticoids (GC) generally appears to act as a 'brake' on aggressive behaviour (Haller & Kruk 2003). For example, animals exposed to prolonged stressors display chronically elevated circulating corticosterone concentrations and decreased aggression, whereas animals with GC hypofunction display pathologically high levels of aggression (Haller & Kruk 2003). In contrast, acute activation of the HPA axis can increase aggression in rodents; e.g. stimulation of hypothalamic brain regions can evoke aggression in addition to rapid activation of the HPA axis (Kruk et al 1998). The switch point between acute and chronic HPA activation remains unspecified. Because the relationship between NO and the HPA axis and the relationship between the HPA axis on aggressive behaviour, it seems reasonable to suggest that NO may be involved in mediating aggression via the HPA axis.

Despite much research, the role of NO in the regulation of the HPA axis remains unspecified (Givalois-Llort et al 2002). NOS is present in discrete hypothalamic areas (i.e. supraopic nucleus [SON] and paraventricular nucleus [PVN]) that regulate neuroendocrine responses (Huang et al 1993) and stimuli that affect pituitary hormone release (i.e. stress, food deprivation, gonadectomy, exposure to endotoxin) can up-regulate nNOS expression. HPA activity is regulated primarily by the actions of the hypothalamic peptide corticotropin-releasing hormone (CRH), which acts on the pituitary to trigger the release of the tropic hormone adrenocorticotrophic hormone (ACTH). ACTH in turn, regulates the release of glucocorticoids (GCs) from the adrenal cortex. Numerous factors act at the level of hypothalamus or pituitary to affect the release of CRH or ACTH, respectively. For example, the cytokines interleukin 1 (IL1), IL1β, IL2 and IL2β increase CRH release both *in vitro* and *in vivo* (McCann et al 2000). *In vivo* treatment with IL1β increases hypothalamic CRH release, as well as plasma ACTH and corticosterone concentrations; IL1β-induced activation of the HPA can be attenuated, however, by pre-treatment with the NOS inhibitor L-NAME (Rivier & Shen 1994).

In addition to the effects of cytokines on HPA activity, several neurotransmitters/neurohormones (e.g. acetylcholine, norepinephrine, prostaglandins) affect CRH secretion and NO has been implicated as a potential neuroendocrine mediator (Nelson et al 1997). NO also appears to play an important role in ACTH release from the pituitary. For example, i.c.v. or peripheral injections of the non-specific NOS inhibitor, L-NAME, attenuate stress-induced ACTH release (Kim & Rivier 2000). In contrast, endotoxin-induced increases in plasma ACTH and corticosterone are enhanced by central pre-treatment with L-NAME (Rivier & Shen 1994, Harada et al 1999). Despite the contradictory results regarding the specific effects of NO on HPA activity, NO exerts a significant role at the level of the HPA, and NO may modulate CRH release differentially depending on whether the stressor is extrinsic or intrinsic (Bilbo et al 2003).

The effects of restraint on antigen-specific delayed-type-hypersensitivity (DTH) responses in the skin are blunted in nNOS$^{-/-}$ mice, and they lost less body mass after stress than WT mice (Bilbo et al 2003). Neuronal NO appears to be involved in the neuroendocrine–immune response to stress, perhaps via glucocorticoid regulation. Despite elevated corticosterone concentrations, nNOS knockout mice appear less 'anxious' or 'fearful' than WT mice, which may contribute to their aggressiveness: e.g. male nNOS$^{-/-}$ mice spend more time in the centre of the open field than WT mice (Bilbo et al 2003).

Pain and aggression

The relationship between pain and aggression appears to be bi-directional; pain often increases aggression (Matto et al 1998, Bigi et al 1993) and aggressive encounters often alter pain perception (Kulling et al 1988, Vivian & Miczek 1999). Alterations in neurochemical expression at many levels of this pathway could result in contextually inappropriate aggressive behaviour, accompanied by altered pain perception. nNOS knockout mice display increased sensitivity to painful stimuli on a hotplate, which may also prolong aggressive interactions (M. Gatien and R. Nelson, unpublished observations).

Depression and aggression

Elevated aggression is often observed with depression (Harvey et al 2003). Growing evidence suggests that NO might mediate the link between aggression and depression (McLeod et al 2001). In humans, postmortem comparisons of brains of depressed people reveal less nNOS staining in the SCN and PVN than in brains from age-matched people who had no history of psychiatric or neurological diseases (Bernstein et al 2002). Blood NO concentrations were decreased in patients with major depression (Selley 2004). NOS inhibitors possess antidepressant- and anxiolytic-like properties in animal models. The selective nNOS inhibitors 7-NI and 1-(2-trifluoromethylphenyl) imidazole (TRIM) decreased the immobility time in the forced swimming test (Volke et al 2003). The magnitude of the effect mimicked that of the tricyclic antidepressant imipramine (Volke et al 2003).

Male nNOS$^{-/-}$ mice possess reduced serotonin turnover rates and deficient serotonin 5-HT$_{1A}$ and 5-HT$_{1B}$ receptors function in several brain areas regulating emotion, including the hypothalamus, hippocampus and amygdala (Chiavegatto et al 2001). Serotonin dysfunction may be responsible for the observed increases in aggression and impulsivity in these mice (Chiavegatto et al 2001). Interestingly, patients exhibiting depressive symptoms, which may include hyperaggression and impulsivity, often possess serotonin dysfunction, as well as significantly elevated cortisol concentrations and decreased negative feedback mechanisms (Harris et al 2000, Tafet et al 2001a,b).

Despite the reduced 5-HT turnover, $nNOS^{-/-}$ mice display fewer 'depressive-like' responses in the Porsolt swim test. That is, $nNOS^{-/-}$ mice swim longer and display fewer 'float times' than WT mice. This may reflect their general preservation in all behaviours, including fighting and mating, rather than a manic state.

Conclusions

Aggressive behaviour is not a unitary process, but is the result of complex interactions among several physiological, motivational and behavioural systems, with contributions from the social and physical environment. The multiple, and often unanticipated, effects of targeted gene disruption on aggressive behaviour must be considered when phenotyping a gene manipulation. Although many other molecules can affect aggressive behaviour (Nelson & Chiavegatto 2001), most agents likely influence aggression via the signalling properties of 5-HT, or possibly GABA.

Interactions between NO and the HPA axis, as well as between NO and 5-HT mechanisms have been implicated in aggressive behaviour. Future studies must consider the environmental (social and physical), hormonal and cellular contributions to aggressive behaviour to understand the molecular mechanisms. Hypothetical psychological constructs such as fear, hunger, anxiety and depression also affect the molecular mechanisms of aggression. A variety of subtle adjustments in 5-HT concentrations, turnover and metabolism, or slight changes in receptor subtype activation, density and binding affinity alone or in combination, and possibly mediated by NO, can influence aggression in different ways by affecting inputs into aggression circuitries. Because aggressive behaviour is not a unitary process, it is likely that multiple changes in 5-HT signalling are associated with different types of aggression. The integrity of this complex pathway seems necessary for normal expression and termination of aggressive behaviour (Chiavegatto & Nelson 2003).

Future studies using gene arrays, inducible gene knockouts and knock-ins, and RNA silencing techniques may be necessary to untangle the multiple influences of various molecules on aggressive behaviour. Multiple levels of analysis, as well as comparative research approaches are necessary to completely reveal the contributions of NO to the molecular bases of aggressive behaviour.

Acknowledgements

I thank Zachary Weil and Brenda Reader for bibliographic assistance, and Dr A. C. DeVries for valuable comments. Support for the studies reported from our laboratory and for writing this review was provided by NIH Grants MH 57535 and MH57760.

References

Azcarate CL 1975 Minor tranquilizers in the treatment of aggression. J Nerv Ment Dis 160: 100–107

Albert DJ, Walsh ML 1984 Neural systems and the inhibitory modulation of agonistic behavior: A comparison of mammalian species. Neurosci Biobehav Rev 8:5–24

Bernstein HG, Heinemann A, Krell D et al 2002 Further immunohistochemical evidence for impaired NO signalling in the hypothalamus of depressed patients. Ann NY Acad Sci 973:91–93

Bigi S, De Acetis L, De Simone R, Aloe L, Alleva E 1993 Neonatal capsaicin exposure affects isolation-induced aggressive behavior and hypothalamic substance P levels of adult male mice (Mus musculus). Behav Neurosci 107:363–369

Bilbo SD, Hotchkiss AK, Chiavegatto S, Nelson RJ 2003 Blunted stress responses in delayed type hypersensitivity in mice lacking the neuronal isoform of nitric oxide synthase. J Neuroimmunol 140:41–48

Blanchard DC, Griebel G, Blanchard RJ 2001 Mouse defensive behaviors: pharmacological and behavioral assays for anxiety and panic. Neurosci Biobehav Rev 25:205–218

Chiavegatto S, Nelson RJ 2003 Interaction of nitric oxide and serotonin in aggressive behavior. Horm Behav 44:233–241

Chiavegatto S, Dawson VL, Mamounas LA, Koliatsos E, Dawson TM, Nelson RJ 2001 Brain serotonin dysfunction accounts for aggression in male mice lacking neuronal nitric oxide synthase Proc Natl Acad Sci USA 98:1277–1281

Citrome L, Volavka J 1997a Psychopharmacology of violence: Part I: assessment and acute treatment. Psychiatr Ann 27:691–695

Citrome, L, Volavka J 1997b Psychopharmacology of violence: Part II: beyond the acute episode. Psychiatr Ann 10:696–703

Connor DF, Boone RT, Steingard RJ, Lopez ID, Melloni RH 2003 Psychopharmacology and aggression: II. A meta-analysis of nonstimulant medication effects on overt aggression-related behaviors in youth with SED. J Emot Behav Disord 11:157–168

Crawley JN 2000 What's wrong with my mouse? Wiley, New York

Dawson TM & Snyder SH 1994 Gases as biological messengers: Nitric oxide and carbon monoxide in the brain. J Neurosci 14:5147–5159

Dawson TM, Dawson VL 1996 Nitric oxide synthase: role as a transmitter/mediator in the brain and endocrine system. Ann Rev Med 47:219–245

De Almeida RM, Nikulina EM, Faccidomo S, Fish EW, Miczek KA, 2001 Zolmitriptan — a 5-HT1B/D agonist, alcohol, and aggression in mice. Psychopharmacology (Berl) 157: 131–141

Demas GE, Eliasson MJL, Dawson TM et al 1997 Inhibition of neuronal nitric oxide synthase increases aggressive behavior in mice. Mol Med 3:611–617

Demas GE, Kriegsfeld LJ, Blackshaw S et al 1999 Elimination of aggressive behavior in male mice lacking endothelial nitric oxide synthase. J Neurosci 19:RC30

Gammie SC, Nelson RJ 1999 Maternal aggression is reduced in neuronal nitric oxide synthase-deficient mice. J Neurosci 19:8027–8035

Gammie SC, Olaghere-da Silva UB, Nelson RJ 2000 3-Bromo-7-nitroindazole, a neuronal nitric oxide synthase inhibitor, impairs maternal aggression and citrulline immunoreactivity in prairie voles. Brain Res 870:80–86

Givalois-Llort, Li S, Pelletier G 2002 Central nitric oxide regulation of the hypothalamic-pituitary-adrenocortical axis in adult male rats. Brain Res Mol Brain Res 102:1–8

Harada S, Imaki T, Chikada N, Naruse M, Demura H 1999 Distinct distribution and time-course changes in neuronal nitric oxide synthase and inducible NOS in the paraventricular nucleus following lipopolysaccharide injection. Brain Res 821:322–332

Haller J, Kruk MR 2003 Neuroendocrine stress responses and aggression. In: Mattson MP (ed) Neurobiology of aggression: understanding and preventing violence. Humana Press, New Jersey, p 93–118

Harris TO, Borsanyi S, Messari S et al 2000 Morning cortisol as a risk factor for subsequent major depressive disorder in adult women. Br J Psychiatry 177:505–510

Harvey BH, McEwen BS, Stein DJ 2003 Neurobiology of antidepressant withdrawal: implications for the longitudinal outcome of depression. Biol Psychiatry 54:1105–1117

Huang PL, Dawson TM, Bredt DS, Snyder SH, Fishman MC 1993 Targeted disruption of the neuronal nitric oxide synthase gene. Cell 75:1273–1286

Huang PL, Huang Z, Mashimo H et al 1995 Hypertension in mice lacking the gene for endothelial nitric oxide synthase. Nature 377:239–242

Ignarro LJ 1990 Biosynthesis and metabolism of endothelium-derived nitric oxide. Ann Rev Pharmacol Toxicol 30:535–560

Jasnow AM, Huhman KL, Bartness TJ, Demas GE 2000 Short-day increases in aggression are inversely related to circulating testosterone concentrations in male Siberian hamsters (Phodopus sungorus). Horm Behav 38:102–110

Kim CK, Rivier CL 2000 Nitric oxide and carbon monoxide have a stimulatory role in the hypothalamic-pituitary-adrenal response to physico-emotional stressors in rats. Endocrinology 141:2244–2253

Kriegsfeld LJ, Dawson TM, Dawson VL, Nelson RJ, Snyder SH 1997 Aggressive behavior in male mice lacking the gene for nNOS is testosterone-dependent. Brain Res 769:66–70

Kruk MR, Westphal KG, Van Erp AM et al 1998 The hypothalamus: cross-roads of endocrine and behavioural regulation in grooming and aggression. Neurosci Biobehav Rev 23:163–177

Kulling P, Frischknecht HR, Pasi A, Waser PG, Siegfried B 1988 Social conflict-induced changes in nociception and beta-endorphin-like immunoreactivity in pituitary and discrete brain areas of C57BL/6 and DBA/2 mice. Brain Res 450:237–246

Le Roy I, Pothion S, Mortaud S et al 2000 Loss of aggression, after transfer onto a C57BL/6J background, in mice carrying a targeted disruption of the neuronal nitric oxide synthase gene. Behav Genet 30:367–373

Manuck SB, Flory JD, McCaffery JM, Matthews KA, Mann JJ, Muldoon MF 1998 Aggression, impulsivity, and central nervous system serotonergic responsivity in a nonpatient sample. Neuropsychopharmacology 19:287–299

Matto V, Skrebuhhova T, Allikmets L 1998 The effect of antidepressants on rat aggressive behavior in the electric footshock and apomorphine-induced aggressiveness paradigms. Methods Find Exp Clin Pharmacol 20:329–337

McCann SM, Kimura M, Karanth S, Yu WH, Mastronardi CA, Rettori V 2000 The mechanism of action of cytokines to control the release of hypothalamic and pituitary hormones in infection. Ann NY Acad Sci 917:4–18

McLeod TM, Lopez-Figueroa AL, Lopez-Figueroa MO 2001 Nitric oxide, stress, and depression. Psychopharmacol Bull 35:24–41

Miczek KA, De Almeida RM 2001 Oral drug self-administration in the home cage of mice: alcohol-heightened aggression and inhibition by the 5-HT1B agonist anpirtoline. Psychopharmacology (Berl) 157:421–429

Miczek KA, Fish EW 2005 Monoamines, GABA, glutamate, and aggression. In: Nelson RJ (ed) Biology of aggression. Oxford University Press, p 114–149

Miczek KA, Maxson SC, Fish EW, Faccidomo S 2001 Aggressive behavioral phenotypes in mice. Behav Brain Res 125:167–181

Miczek KA, Fish EW, De Bold JF 2003 Neurosteroids, GABAA receptors, and escalated aggressive behavior. Horm. Behav 44:242–257

Mineur YS, Prasol DJ, Belzung C, Crusio WE 2003 Agonistic behavior and unpredictable chronic mild stress in mice. Behav Genet 33:513–519

Moncada S, Higgs A 1993 The L-arginine-nitric oxide pathway. N Engl J Med 329:2002–2012
Moyer KE 1971 Physiology of hostility. Markum, Chicago
Nathan C 1992 Nitric oxide as a secretory product of mammalian cells. FASEB J 6: 3051–3064
Nathan C, Xie Q-W 1994 Nitric oxide synthases: roles, tolls, and controls. Cell 78:915–918
Nelson RJ 1997 The use of genetic "knock-out" mice in behavioral research. Horm Behav 31:188–196
Nelson RJ, Chiavegatto S 2001 Molecular basis of aggression. Trends Neurosci 24:713–719
Nelson RJ, Demas GE, Huang PL et al 1995 Behavioral abnormalities in male mice lacking neuronal nitric oxide synthase. Nature 378:383–386
Nelson RJ, Kriegsfeld LJ, Dawson VL, Dawson TM 1997 Effects of nitric oxide on neuroendocrine function and behavior. Front Neuroendocrinol 18:463–491
New AS, Buchsbaum MS, Hazlett EA et al 2004 Fluoxetine increases relative metabolic rate in prefrontal cortex in impulsive aggression. Psychopharmacology 176:451–458
Nyberg JM, Vekovischeva O, Sandnabba NK 2003 Anxiety profiles of mice selectively bred for intermale aggression. Behav Genet 33:503–511
Olivier B, Mos J, van Oorschot R, Hen R 1995 Serotonin receptors and animal models of aggressive behavior. Pharmacopsychiatry 28(Suppl 2):80–90
Rivier C, Shen GH 1994 In the rat, endogenous nitric oxide modulates the response of the hypothalamic-pituitary-adrenal axis to interleukin-1 beta, vasopressin and oxytocin. J Neurosci 14:1985–1993
Selley ML 2004 Increased (E)-4-hydroxy-2-nonenal and asymmetric dimethylarginine concentrations and decreased nitric oxide concentrations in the plasma of patients with major depression. J Affect Dis 80:249–256
Simon NG 2002 Hormonal systems regulating the development and expression of aggression. In: Pfaff D, Etgen A, Rubin R, Fahrbach B (eds) Hormones, brain and behavior. Academic Press, San Diego, p 39–85
Singh R, Pervin S, Shryne J, Gorski R, Chaudhuri G 2000 Castration increases and androgens decrease nitric oxide synthase activity in the brain: physiologic implications. Proc Natl Acad Sci USA 97:3672–3677
Soma KK, Wingfield JC 2001 Dehydroepiandrosterone in songbird plasma: seasonal regulation and relationship to territorial aggression. Gen Comp Endocrinol 23:144–155
Spigset O 1999 Adverse reactions of selective serotonin reuptake inhibitors: reports from a spontaneous reporting system. Drug Saf 20:277–287
Svare B 1983 Hormones and aggressive behavior. Plenum, New York
Swann AC 2003 Neuroreceptor mechanisms of aggression and its treatment. J Clin Psychiatry 64:26–35
Tafet GE, Toister-Achituv M, Shinitzky M 2001a Enhancement of serotonin uptake by cortisol: a possible link between stress and depression. Cogn Affect Behav Neurosci 1: 96–104
Tafet GE, Idoyaga-Vargas VP, Abulafia DP et al 2001b Correlation between cortisol level and serotonin uptake in patients with chronic stress and depression. Cogn Affect Behav Neurosci 1:388–393
Vincent SR, Kimura HJ 1992 Histochemical mapping of nitric oxide synthase in the rat brain. Neuroscience 4:755–784
Vivian JA, Miczek KA 1999 Interactions between social stress and morphine in the periaqueductal gray: effects on affective vocal and reflexive pain responses in rats. Psychopharmacology (Berl) 146:153–161
Volke V, Wegener G, Bourin M, Vasar E 2003 Antidepressant- and anxiolytic-like effects of selective neuronal NOS inhibitor 1-(2-trifluoromethylphenyl)-imidazole in mice. Behav Brain Res 140:141–147

Walsh MT, Dinan TG 2001 Selective serotonin reuptake inhibitors and violence: a review of the available evidence. Acta Psychiatr Scand 104:84–91

Wen JC, Hotchkiss AK, Demas GE, Nelson RJ 2004 Photoperiod affects neuronal nitric oxide synthase and aggressive behaviour in male Siberian hamsters (*Phodopus sungorus*). J Neuroendocrinol 16:916–921

Weiner CP, Lizasoain I, Baylis SA, Knowles RG, Charles IG, Moncada S 1994 Induction of calcium-dependent nitric oxide synthases by sex hormones. Proc Natl Acad Sci USA 91:5212–5216

DISCUSSION

Lovell-Badge: You said that nNOS expression is altered, or the number of neurons expressing it is altered. Which is it? Is nNOS expression being regulated in the hamsters, for example, by circadian rhythms? It the gene being regulated or is it an alteration in the numbers of neurons?

Nelson: We did not examine circadian rhythms in nNOS expression in the Siberian hamsters. It looks like photoperiod alters the number of neurons in these animals. In mice, there was a distinct circadian rhythm in nitric oxide production as evidenced by citrulline production (Kriegsfeld et al 1998). However, circadian rhythms were not disrupted in $nNOS^{-/-}$ mice (Kriegsfeld et al 1999).

Robins: In those hamsters that are released from testosterone regulation, is there a difference in fat content between the responders and the non-responders? Perhaps oestrogen levels are important and fat can make substantial amounts of oestrogen: This could tie into leptin regulation.

Nelson: Yes, that whole constellation of body fat and reproductive response is tied together. You can do some tricks such as yoked feeding, where you can cause the animals to lose weight, but the effect is still seen.

Robins: Are you looking to see how easily heritable the responder/non-responder phenotype is?

Nelson: No. However, Bronson and colleagues could fix the non-responder trait in four generations (Desjardins et al 1986). I believe that Paul Heideman at William and Mary in Virginia has developed two separate lines of responders and non-responders in deer mice (*Peromyscus maniculatus*).

Martinez: What is the relationship between previous experiences of winning and subsequent aggressive behaviour in knockout animals?

Nelson: The nNOS knockout animals start out aggressive and continue to show a high rate of aggressive behaviours in subsequent behavioural testing. The eNOS animals start off docile and become more aggressive with experience.

Olivier: I saw the nNOS mice had high corticosterone levels. As they become older do they develop Cushing syndrome?

Nelson: We haven't seen this yet. They don't live long without lots of other problems: they have enlarged stomachs and bladders. If they go too long they become frail.

Olivier: Is the CRF system in the brain affected?

Nelson: We have only recently begun to look at this.

Olivier: What happens to corticosterone levels in eNOS knockout mice?

Nelson: I'm waiting for funding to address this question! We have not yet assayed corticosterone concentrations.

Brodkin: The effects of group housing on the knockouts with brothers are very interesting. I don't know whether the litter sizes are large enough to try this, but if they were group housed with null homozygotes for NOS, what would happen?

Nelson: That is a good question. One of the experiments I proposed in the application was to do factorial combinations with wild-types and knockouts to try to parse this out.

Koolhaas: I may have an explanation for your observations on eNOS knockouts. It is involved in vasomotion, so temperature regulation might be a good candidate. We do a lot of temperature recording with telemetry, and aggressive behaviours are highly metabolically demanding, resulting in a rapid rise in body temperature. High body temperature is a very strong factor for inhibiting aggressive behaviour. If it gets too high, aggressive behaviour, food intake and sexual behaviour all stop.

Nelson: That is a good suggestion. Another possibility is that the effect on vascular tone could be reflecting altered neuroendocrine signals from the hypothalamus and pituitary could be altered.

Pfaff: It is not a problem, it is a novel mechanism.

Ferris: Where is the highest staining for NOS in the rodent brain?

Nelson: It is throughout the limbic system, and the highest level is in the cerebellum.

Keverne: And in the accessory olfactory bulb.

Ferris: If you were to go any place in the brain and look for the hotspot for initiating aggression, there is compelling evidence across species that the medial basal hypothalamus is the critical area.

Pfaff: mRNA for nNOS is expressed at quite high levels in the ventromedial hypothalamus and it is highly oestrogen sensitive. Oestrogen treatment of an ovariectomized animal significantly elevates the transcript levels (Ceccatelli et al 1996, Rachman et al 1998).

Ferris: Have you looked at vasopressin in your Siberian hamsters?

Nelson: No.

Keverne: One important point you made was this business about disregarding social signals. One thing that nNOS does in the knockouts is that it impairs learning and memory. Therefore, when these animals are maturing they are not going to experience playfighting, and thereby learn to respond to the signals of 'giving up', and hence develop the appropriate behaviours when they become adults. When you knocked out the effect of nNOS with a drug in adults, you showed that they were the first to attack. But were they as systematically aggressive, and did they ignore the signals of defeat?

Nelson: That is a good question. The amount of the aggression was the same as with the knockouts but the quality wasn't as vicious. They did not ignore the social cues to the extent that the nNOS knockouts did. For example, when an opponent showed submissive behaviour they tended to stop fighting.

Keverne: The second point concerns social cues: since it is so highly expressed in the accessory olfactory system, is it possible that it could be having some impact there?

Nelson: Very likely it is. We haven't gone through a systematic examination of this. They can find hidden food so the olfactory system works. We used Q tips swabbed with vaginal secretions from an oestrous versus non-oestrous mouse, and they could discriminate on this basis.

C Blanchard: In your female maternal aggression task, was the intruder a male or female?

Nelson: We did all permutations. They are mostly aggressive towards the males. The NOS knockouts behaved similarly.

C Blanchard: I remember some work by Parmigiani and his group (Parmigiani et al 1989) using the serenic fluprazine, which virtually abolished many types of male–male offensive aggression. In females in a maternal aggression situation it severely reduced aggression of females towards intruding females, but it did little to the aggression of females towards intruding males. This suggests that fluprazine may be able to differentiate between offensive and defensive aggression.

Olivier: This only holds in mice and not rats.

Gammie: The simplest explanation now for the knockout effects on maternal aggression is with the finding of elevated CRH. We have shown (Gammie et al 2004) that we can dose-dependently inhibit maternal aggression in mice without affecting other maternal behaviour. We have also mapped brain regions that are activated by CRH.

Pfaff: When there are differences in aggression according to the identity or gender of the intruder, how rapidly are the differences expressed? And what does this mean about the likely participation of the vomeronasal organ, which I think of as a temporally sluggish communication device? When we put our ER knockout females in the cage of a male mouse, the attack is immediate—the female is treated like a male.

Nelson: The nNOS knockouts were very slow at defending their pups. They were good mothers in other regards except they didn't defend against the male intruder. When we examined the wild-type females and stained the brains for citrulline we were able to determine where NO was produced.

Immunohistochemical assessment of citrulline provides an indirect 'snapshot' of NO release during a specific behaviour. NO is the breakdown product when citrulline is cleaved enzymatically from arginine by NOS and can be analysed chemically and immunohistochemically as an indirect measurement of NO production. We observed a significant increase in the number of citrulline-positive cells in the medial preoptic nucleus, the suprachiasmatic nucleus, and the subparaventricular zone regions of the hypothalamus in aggressive lactating females relative to control mice (Gammie & Nelson 1999). In other regions of the brain, no changes in the number of citrulline-positive cells were observed across either groups or treatments. We also conducted a comparative study with prairie voles, which are biparental and the males also provide nest defence. In this case we observed staining for NOS in the same brain areas of the females (Gammie & Nelson 2000). These results provide two indirect lines of evidence that NO release is associated with maternal aggression.

Brembs: I have a question about the comparison between the NO knockout mice and the pharmacologically treated mice. Did you look at the serotonin turnover in both?

Nelson: That is on my list of things to do. It is tricky using 7-NI, because in a dish NO production is knocked down about 85% of normal production, and then it starts coming back. As a result we have to give this drug every 6 h, which makes it a challenging project (Demas et al 1997). There is a new class of drugs that only have to be given once a day so we are trying those now.

Pfaff: Is there a cell biology of NO in the brain which could give us hints about how serotonin turnover would be reduced?

Nelson: No, there isn't. It hasn't been worked out yet.

Suomi: I was intrigued by your response to Gary's question, comparing the aggression in the knockouts with those that were pharmacologically induced. You said that although the absolute levels were comparable, the knockouts weren't responsive to submissive cues while the drug-induced ones were. Is that correct?

Nelson: That was my qualitative judgement.

Suomi: Did the drug-induced animals have a history of interacting with other animals before that?

Nelson: Not any more than the knockouts did. They were reared in the same context. I don't think this result was because they had or had not seen these cues before.

Brodkin: I wanted to suggest a topic for general discussion: the relationship between aggressive behaviour and HPA axis functioning. There is the issue of HPA axis functioning in dominance hierarchies in primates, and I have seen some literature about cortisol levels in children who show antisocial behaviour. Several studies have found a correlation between lower levels of cortisol and higher levels of antisocial behaviour in children and adolescents (McBurnett et al 2001, Pajer et al 2001, Shoal et al 2003). So I was surprised and interested in your finding that the NOS knockouts showed higher aggressive behaviour and also had higher baseline ACTH and corticosterone.

Pfaff: That is a good topic for general discussion. There are CRF projections to the locus coeruleus which could make the animal hyper-responsive and twitchy. Aside from the HPA axis itself, CRF has many other activities.

Koolhaas: I was happy to see your data from the forced swimming test. This test is generally used as a test for depression and you had some difficulties understanding why your animals would differ in a test for depression. However, if you adopt the idea that aggression is a form of active coping with environmental problems, it fits perfectly. The concept of proactive and reactive coping would predict that an animal that is highly aggressive in its home territory, i.e. actively deals with a social problem, will be likely to show active swimming and escape in the forced swim test as well. Do you think the concept of aggression as a coping style might help in explaining the behavioural profiles of your mice?

Nelson: I think it could help, but you could explain it the other way.

Koolhas: If you take the extremes of our population of wild-type rats or our selection lines of mice, you would find perfect correlation: the low aggressive rat and low aggressive mouse floats, whereas the high aggressive animals continue to swim and struggle.

Nelson: That's a good idea.

Olivier: In one of your slides you showed effects of serotonin agonists and antagonists on aggression. You mentioned that a 5-HT_{1A} antagonist enhanced aggressive behaviour. Which one?

Nelson: There was a generalized slide showing that in general, animals with elevated serotonin reduced aggression, whereas animals with low serotonin increased aggression. There was another slide that showed that specific 5-HT_1 agonists decreased aggression in the $nNOS^{-/-}$ mice. The 5-HT_{1A} agonist, 8-OH-DPAT, and the 5-HT_{1B} agonist, CP-94,253, significantly decreased aggressive behavior in both wild-type and $nNOS^{-/-}$ mice by reducing the number and duration of attacks and increasing the latency to first attack.

Olivier: The reason I ask is that animals treated with 5-HT_1 antagonist never show a proaggressive phenotype.

Pfaff: The females didn't have a phenotype. This suggests to me that they have limits on aggressive behaviours that the males didn't have, which were independent of NO, or they had a failure of other facilitators of aggression, independent of NO.

Nelson: They reduced maternal aggression and nest defence.

Pfaff: And, do I understand, not of the vicious aggression you showed with the males.

Nelson: This doesn't surprise me too much because they don't normally show territorial behaviour anyway.

Pfaff: Perhaps, the reason you are not surprised is because you have moved the problem to a different level. My question could now be restated at that level: why don't they have the territorial aggression at the beginning?

Nelson: It is because of the way they live in nature, which has selected for their behaviour. Males defend the living space in the demes. I suspect that if we changed the aggression test by food-restricting the females and then have them compete for food, I would then have a different phenotype.

Lesch: I am intrigued by the specificity of the nNOS knockout on the serotonergic function. Is there any evidence for dopaminergic or noradrenergic dysregulation?

Nelson: No. Silvana Chiavegatto, who was a postdoctoral fellow in my group, measured dopamine and by-products. We saw some significant change here and there, but there were not overall patterns and we haven't pursued it. We saw no significant changes with GABA in the brains of the $nNOS^{-/-}$ mice, either.

Lesch: I have a comment regarding the interaction of serotonin and the HPA axis. There is a tight reciprocal regulation of both systems. There is strong expression of glucocorticoid receptors in serotonergic neurons, and there are also 5-HT_{1A} receptors strongly expressed hypothalamic nuclei. So these two systems interact strongly and cross-regulate themselves. It is not surprising that alterations in the serotonergic system cause HPA axis disturbances.

References

Ceccatelli S, Grandison L, Scott RE, Pfaff DW, Kow LM 1996 Estradiol regulation of nitric oxide synthase mRNAs in rat hypothalamus. Neuroendocrinology 64:357–363

Demas GE, Eliasson MJL, Dawson TM et al 1997 Inhibition of neuronal nitric oxide synthase increases aggressive behavior in mice. Mol Med 3:611–617

Desjardins C, Bronson FH, Blank JL 1986 Genetic selection for reproductive photoresponsiveness in deer mice. Nature 322:172–173

Gammie SC, Nelson RJ 1999 Maternal aggression is mediated by nitric oxide. J Neurosci 19:8027–8035

Gammie SC, Nelson RJ 2000 Maternal and mating-induced aggression are associated with elevated citrulline immunoreactivity in the paraventricular nucleus in prairie voles. J Comp Neurol 418:182–192

Gammie SC, Negron A, Newman SM, Rhodes JS 2004 Corticotropin-releasing factor inhibits maternal aggression in mice. Behav Neurosci 118:805–814

Kriegsfeld LJ, Elliason M, Demas GE et al 1998 Nocturnal motor coordination deficits in neuronal nitric oxide synthase knock-out mice parallel altered citrulline production. Neuroscience 89:311–315

Kriegsfeld LJ, Demas GE, Dawson TM, Dawson VL, Lee S, Nelson RJ 1999 Circadian behavior in mice with targeted disruption of the gene for the neuronal isoform of nitric oxide synthase. J Biol Rhythms 14:20–27

McBurnett K, Lahey BB, Rathouz PJ, Loeber R 2001 Low salivary cortisol and persistent aggression in boys referred for disruptive behaviour. Arch Gen Psychiatry 58:513–515

Pajer K, Gardner W, Rubin RT, Perel J, Neal S 2001 Decreased cortisol levels in adolescent girls with conduct disorder. Arch Gen Psychiatry 58:297–302

Parmigiani S, Rodgers RJ, Palanza P, Mainardi M, Brain PF 1989 The inhibitory effects of fluprazine on parental aggression in female mice are dependent upon intruder sex. Physiol Behav 46:455–459

Rachman IM, Unnerstall JR, Pfaff DW, Cohen RS 1998 Estrogen alters behavior and forebrain c-fos expression in ovariectomized rats subjected to the forced swim test. Proc Natl Acad Sci USA 95:13941–13946

Shoal GD, Giancola PR, Kirillova GP 2003 Salivary cortisol, personality, and aggressive behaviour in adolescent boys: a 5-year longitudinal study. J Am Acad Child Adolesc Psychiatry 42:1101–1107

General discussion II

Craig: I would like to raise the general problem of looking at individual genes in systems for human behaviours. This might be called the candidate gene approach. Earlier we asked why sometimes there is one answer, and other times a different answer. There are several reasons for this. The first is the general effect size found in human behavioural studies. We discussed yesterday the number of quantitative trait loci (QTLs) for different behaviours. I think you will be lucky to find an effect size of 2%, so we need quite a lot of power to detect this. The second issue is deciding what functional allele we are looking at. You can look at a promoter variant but it may be in linkage disequilibrium with something else within an intron, which is also functional in processing or something of that sort. The closer we look at most of these genes we find introns that have a lot of VNTRs (variable number of tandem repeats) which probably have some kind of functional significance, so we have to be careful about which alleles we pick up. Then there is the problem of the phenotype. Earlier on we were talking about depression and serotonin transporters, along with anxiety and neuroticism. It is quite clear that these things factor on one another, but we don't know to what extent. You can pick one of them and then pick up some candidate effects, and then another one and pick up another candidate effect. Finally, with respect to serotonin transporter and the monoamine oxidase A (MAOA) studies, it is clear that we are only going to get a good increase in effect size if we factor in life events. The depression study demonstrates that we only find the serotonin transporter low allele effect if there is an abusive childhood. It is not surprising to me, bearing all these factors in mind, that it has proved so difficult to get the results reported so far. I would also like to mention one other study using the serotonin transporter on an adolescent sample. We have just published a paper showing that there is a very good trend for the low activity allele of the serotonin transporter being associated with neuroticism in adolescent girls (Eley et al 2004).

Manuck: We reported recently that the socioeconomic status of individuals covaries with brain serotonergic activity as a function of allelic variation in the common regulatory VNTR of the serotonin transporter gene (5-HTTLPR) (Manuck et al 2004). In particular, we found a blunted serotonergic responsivity associated with lower socioeconomic position among persons homozygous for the 5-HTTLPR short allele, whereas no association appeared in the group that was homozygous for the alternate, long allele; an intermediate (and significant) association was seen among heterozygotes. Perhaps most interestingly, serotonergic responsivity (measured as the rise in plasma prolactin concentration induced by fenfluramine) declined across genotypes—from long–long to short–short homozygotes—among individuals of lower socioeconomic status, but tended to rise over the same range of genotypes in persons of more advantaged social position. One implication of this finding is that genetic associations reported in the literature may vary—and indeed, vary even in the direction of allelic association observed for particular phenotypes—if the populations studied differ systematically in important background characteristics, including aspects of subjects' social environments and developmental experiences.

Brembs: Let me try to make this topic even more general. From what I have learned so far, it mimics a discussion people have had fairly recently in my main research area (learning and memory), where a lot of developmental genes are 'recycled' in the brain. They are used very early in development, and then are re-used in the brain to modify the brain cells during learning. A symposium a few years ago discussed the question, is there a difference between learning and development?

Lesch: Could you give examples of these genes?

Brembs: All the genes that increase the number of spines or change the synapse are involved. PKA, for example, is one that is considered to be an integrator of cellular signals from the conditioned stimulus and the unconditioned stimulus. Then there are the genes involved in the various second messenger cascades, most notably the cAMP cascade or the processes involving the different MAP kinases. Then there are the genes coding for a variety of structurally relevant cell adhesion molecules, such as NCAM or L1. They're all part of the cellular machinery, and every change in a cell, whether it is prompted by a cue that is developmental or by experience, uses this cellular machinery. The same takes place in aggression. As we have discussed extensively, aggression has implications for a range of different systems. Other circuits have interactions with circuits mediating aggression, which leads to knockouts showing opposite phenotypes in males and females, or pharmacology not being able to reproduce the knockouts, and all kinds of diverging effects from the manipulations that we carry out.

What I am saying may either be considered heretic or trivial, depending on your standpoint. But it seems that molecular studies on a system that is so highly interactive with other systems, may work just as a coarse first approach, just blindly probing to see whether a gene has any effect, and if so where. Then what one would like to do to get functional models would be to manipulate neuron firing rates and activity, rather than the molecules making up those and many other neurons (or the neurotransmitters/-modulators common to those and many other neurons). The latest fashion in learning and memory research in

Drosophila is to use genetic constructs that switch off certain neural pathways during specific phases of the experiment.

Pfaff: Because of the pleiotropy of a given gene's actions in the CNS, knockout studies are relatively blunt tools. On the basis of the success evidenced in several of the papers here in this symposium, new studies are successfully approaching those molecular mechanisms actually associated with the electrical activity of the relevant neurons at 'time t', just before aggression begins (or doesn't begin). These will be more incisive and more accurate. Is this a correct inference from your discussion?

Brembs: Yes.

Koolhaas: You mentioned structural genes. In a microarray study in our selected short and long attack latency mice, the genes that differed mostly were all structural genes, much to our surprise. This supports what you have said.

Martinez: I want to discuss the problem of the phenotype, which has been raised. In animal studies, I have been working on aggression for more than 20 years, and the problems have remained the same: the problems of the phenotype and how we measure aggression. Perhaps it is time to write a guide for those who are going to start research on aggression in animals, giving detailed information about settings, opponents, timing and measurement of aggressive behaviour and also the whole spectrum of behaviours shown by animals during the aggressive interaction.

Dulac: Such guides exist.

Manuck: In psychology, underlying structures of complex phenotypes are often elucidated by applying factor analytic procedures to multiple observations (or measurements) that are thought to tap aspects of a common construct. With respect to aggression, such studies have generated several instruments—questionnaires and semi-structured interviews—that yield both aggregated indices of aggressive disposition and measures having subscales reflecting specific features of aggression (e.g. physical assault, verbal hostility, indirect aggression and irritability, inhibited anger, etc.). Is there anything similar in the animal literature?

Dulac: There is a review article on animals analysed in different laboratories, and the discrepancy is striking (Wahlsten et al 2003).

Martinez: In my thesis, it was very simple. I looked at the effects of antiandrogens on intermale aggression in mice. Contradictory results had been reported in the literature. Then, in my study, I used the different types of opponents used in these previous studies and I found all the contradictory results reported.

References

Eley TC, Sugden K, Corsico A et al 2004 Gene–environment interaction analysis of serotonin system markers with adolescent depression. Molecular Psychiatry 9:908–915

Manuck SB, Flory JD, Ferrell RE, Muldoon MF 2004 Socio-economic status covaries with central nervous system serotonergic responsivity as a function of allelic variation in the serotonin transporter gene-linked polymorphic region. Psychoneuroendocrinology 29: 651–668

Wahlsten D, Metten P, Phillips TJ et al 2003 Different data from different labs: lessons from studies of gene–environment interaction. J Neurobiol 54:283–311

Serotonergic mechanisms in aggression

Berend Olivier

Department of Psychopharmacology, Utrecht Institute of Pharmaceutical Sciences and Rudolf Magnus Institute for Neuroscience, Faculty of Pharmaceutical Sciences, Utrecht University, Sorbonnelaan 36, 3584CA Utrecht, The Netherlands, and Yale University School of Medicine, Department of Psychiatry, New Haven, USA

Abstract. The serotonergic system in the CNS has complex interactions with many, if not all other neurotransmitter systems in the brain. Its localization, distribution and amazing receptor diversity makes it an appealing system for modulatory aspects in many basic behaviours, including food and water intake, sexual behaviour and aggression. Notwithstanding decades of research into the putative role of the serotonergic system in aggression, no clear picture about its specific role has emerged. It seems, dependent on state or trait, to be involved in either the performance or the termination of aggressive behaviours. The present technology appears not developed enough to give answers to these questions. Application of drugs and particular selective ligands for certain subtype receptors seems a more promising approach to unravelling the role of 5-HT in aggression. The (postsynaptic) 5-HT_{1B} and to a lesser extent, the 5-HT_{1A} receptor seems to play a prominent role, at least in rodents, in the modulation of (offensive) aggression.

2005 Molecular mechanisms influencing aggressive behaviours. Wiley, Chichester (Novartis Foundation Symposium 268) p 171–189

The serotonergic system and 5-HT receptors

The serotonin (5-HT) system in the CNS contains a limited, but well-defined number of serotonergic cells. The cell bodies (soma) are mainly located in the mid- and hindbrain (Törk 1990) and serotonergic neurons project both to rostral and caudal areas of the brain (Jacobs & Azmitia 1992). In particular, it is thought that the rostral projections play a big role in the involvement of the serotonergic system in the pathology of various psychiatric disorders.

The serotonergic system is complex and, in the last decade, an enormous volume of new findings have dramatically changed the simple concept of the neuron-neurotransmitter-receptor axis. At present, 14 different serotonin receptors can be distinguished within the serotonin receptor family: 5-HT_{1A}, 5-HT_{1B}, 5-HT_{1D}, 5-ht_{1E}, 5-ht_{1F}, 5-HT_{2A}, 5-HT_{2B}, 5-HT_{2C}, 5-HT_3, 5-HT_4, 5-ht_{5A}, 5-ht_{5B}, 5-HT_6 and 5-HT_7.

The various receptors are neuro-anatomically localized at different sites in the CNS (Palacios et al 1990). Thus, 5-HT_{1A} receptors are abundant in the hippocampus, septum, neocortex and raphe nuclei; $5\text{-HT}_{1B/1D}$ receptors show high density in the pallidum and substantia nigra; and 5-HT_{2C} receptors have a high occurrence in the hypothalamus, pallidum, substantia nigra and, particularly, in the choroid plexus. 5-HT_{2A} receptors are mainly localized in the neocortex, whilst 5-HT_3 receptors are common in the substantia gelatinosa. The distribution of 5-HT_4 receptors is widespread in the striatum, olfactory tubercle, nucleus accumbens, globus pallidus and substantia nigra (Grossman et al 1993) whilst 5-HT_5 receptors are predominantly located in the cerebral cortex, hippocampus, habenula, olfactory bulb and cerebellum (Plassat et al 1992). 5-HT_6 receptors are present in the brain and 5-HT_6 mRNA is strongly expressed in olfactory tubercles, striatum, nucleus accumbens, hippocampus, olfactory bulb and cerebral cortex (Grailhe et al 1997). Finally, 5-HT_7 receptors are present in the hippocampus, cerebral cortex, hypothalamus, thalamus and amygdala (Grailhe et al 1997).

5-HT_{1A} receptors are localized presynaptically on the cell bodies and dendrites (so called somatodendritically) of 5-HT neurons in the raphe nuclei and postsynaptically on many non-serotonergic neurons. 5-HT_{1B} (and 5-HT_{1D}) receptors are also localized pre- and postsynaptically; presynaptically, 5-HT_{1B} receptors are localized as so-called autoreceptors on the axon terminals (Zifa & Fillion 1992), whilst postsynaptic 5-HT_{1B} receptors are heteroreceptors localized at the axon terminals of non-serotonergic neurons (Boschert et al 1994). All other 5-HT receptors are presumably localized postsynaptically (Bonaventura et al 1998).

Besides all these receptors the serotonergic transporter (5-HTT) plays an important role in the modulation of serotonergic neurotransmission. 5-HTT is localized both at the terminal portion of the axon and at the cell body of the 5-HT neuron (Hoffman 1993, D'Amato et al 1987, Hrdina et al 1990, Chen et al 1992).

The 5-HT receptor family is part of two extended gene superfamilies; the G protein-coupled receptor superfamily and the ligand-gated ion channel superfamily. The $5\text{-HT}_{1,2,4,5,6}$ and $_7$ receptors are linked to the modulation of either adenylate cyclase or phosphoinositol turnover via G proteins, whereas 5-HT_3 receptors modulate ion channels.

The activity of a serotonergic neuron is presumably regulated via two kind of autoreceptors (5-HT_{1A} and $_{1B}$) and the 5-HT transporter (Piñeyro & Blier 1999). When the neuron fires, the neurotransmitter is released from its terminals and activates for some time all available 5-HT receptors. In order to regulate the firing and the release of serotonin, several mechanisms are around to modulate the firing of the 5-HT neuron. First, 5-HT-transporters in the synaptic terminals, but also at the cell bodies and the dendrites of the 5-HT neurons, bring 5-HT back into the neuron via an uptake mechanism. This process, 5-HT reuptake, is a very important mechanism of a cell to restore its resting condition in order to be able to fire again and to avoid over stimulation of receptors. A second mechanism contributing to ceasing cell firing and stopping release, is activation of the 5-HT_{1B} autoreceptor at the level of the synaptic terminals, leading to a direct inhibition of 5-HT release. A third mechanism is constituted by the somatodendritic 5-HT_{1A} autoreceptors, which, upon activation, directly inhibit cell firing and, consequently, serotonin release. Whether the endogenous serotonin necessary for this inhibition derives from release of 5-HT from the somatodendritic areas themselves or originates from terminals from neighbouring cell in the raphe nuclei is not completely clear yet (Piñeyro & Blier 1999). The interplay of these three processes leads to an apparently highly fine-tuned system of firing patterns of the serotonergic neurons, needed to modulate the various and extensive functions this neurotransmitter is involved in.

The 5-HT_{1B} receptor, previously called 5-HT_{1B} in rodents and 5-$HT_{1D\beta}$ in other species, including human, is now called r5-HT_{1B} and h5-HT_{1B}, respectively (Hartig et al 1996). The 5-HT_{1B} receptors from different species have a large similarity and corresponding functions, but the pharmacology displayed by the rodent and non-rodent versions is quite different, although this difference is only due to one amino acid in one transmembrane domain of the seven-transmembrane TM receptor. No functional differences in 5-HT_{1B} autoreceptors and heteroreceptors have been reported, but some genetic variation in the h5-HT_{1B} receptors has been found (Gothert et al 1998); a mutation in the third transmembrane domain has an allele frequency of 2% and leads to a receptor with a deviant pharmacology from the wild-type (98% allele frequency). The functional consequences of these kind of genetic variations are unknown, but potentially could cause disturbances in CNS functioning.

Is serotonin inhibitory in aggression?

The big dogma in the relationship between serotonin and aggression is that 5-HT inhibits aggression, mainly derived from studies in which serotonin levels in the brain were decreased by neurotoxic agents like pCPA or 5,7-DHT (5,7-dihydroxytryptophan), that deplete serotonin from serotonergic cells. Such an inverse relationship between 5-HT and aggression has been found in animals and humans, although in the latter measurements on 5-HT activity were based on CSF levels of the main metabolite of serotonin, 5-hydroxyindoleacetic acid (5-HIAA). Notwithstanding severe criticisms on this parameter, for many years it was the only measure in humans reflecting (indirectly) the functional status of the 5-HT system. In animals, 5-HT and its metabolite 5-HIAA can be measured directly in the brain and it could be assumed that the inverse relationship between functional serotonergic activity and aggression should easily be established. However, several contradictory results have been found and even reports of a positive relationship between 5-HT and aggression occur. In humans aggression is associated with suicidal behaviour and both seem to be associated with low serotonergic function, although it is possible that both phenomena are independently regulated (for an extensive discussion on the neurobiology of suicidal behaviour, see Mann 2003).

Measurement of contents of 5-HT and 5-HIAA in postmortem brain tissue and determining a turnover rate from these two parameters was originally described to be lower in aggressive vs. non-aggressive mice (Giacalone et al 1968). Successively measuring CSF samples in humans more or less supported this serotonergic hypofunction (Brown et al 1979, Linnoila et al 1983, Kruesi et al 1990). However, this 5-HT hypofunction or deficiency trait more recently has been associated with impulsivity and risk-taking behaviour rather than aggression *per se* (Mann 2003). A causal relation between 5-HT activity and aggression or impulsivity cannot be derived from static measurements of 5-HT or 5-HIAA measurements in brain tissue or CSF. A functional role of serotonergic neurons in the initiation, execution and stopping of aggression (Coccaro 1989, Miczek et al 2002) still has to be established although some progress has been made using *in vivo* microdialysis techniques in freely moving (aggressive) animals. This technique however, still lacks sufficient resolution because sample time (minutes) is still of a different magnitude than the actual behaviour (seconds). Van Erp & Miczek (2000) measured extracellular serotonin (and dopamine) release in 10 min samples in the nucleus accumbens (NAc) and prefrontal cortex (PFC) in rats, before, during and after a 10 min aggressive interaction with a male conspecific. During the agonistic interaction no detectable change in 5-HT release was found in the NAc, but 5-HT levels were already decreased during fighting in the prefrontal cortex. After the confrontation 5-HT levels in the PFC remained lowered (compared to pre-confrontation baseline) for at least one hour, whereas 5-HT in the NAc was not affected. Dopamine levels were enhanced in both brain areas after (but not during) the agonistic confrontation. However, 5-HT levels were decreased in the NAc of rats that have been conditioned to fight at a specific time each day over a 10-day period (Ferrari et al 2003). In the latter experiments heart rate and dopamine release were concurrently measured and both were raised in anticipation of the fight. Apparently, the actual performance of aggression can be dissociated from the anticipation of a fight, where dopamine plays an important role in the physiological and behavioural sequels around the performance and anticipation of aggression, whereas serotonin seems to be particularly related to termination of aggression.

Measuring electrophysiological events happening in the serotonergic neurons during the performance of aggressive behaviour would be very helpful in unravelling the precise role of the serotonergic system but this seems technically not yet feasible. Moreover, the serotonergic system is not made up of one homogeneous mass of cells but is differentiated; the dorsal and median raphe nuclei projecting to the different areas of the forebrain are the most prominent sources of serotonergic neurons innervating areas involved in the initiation, performance and termination of aggressive behaviour. Interestingly, no systematic studies have so far been performed that attempt to delineate the role of the different serotonergic cell groups in various aspects of aggression, although local lesions or local application of drugs in the dorsal or median raphe nuclei have been performed. It is highly unlikely that all serotonergic cell groups are involved and selective blockade or activation of individual cell groups in determining its role in aggression would be very fruitful.

A recent approach to unravelling the role of the 5-HT system in aggression is studying the differences between highly aggressive and low-aggressive individuals as has recently been pursued by the group of Koolhaas (De Boer et al 2003). They argued, based on the assumption that the individual level of aggression of a rat (offensive aggression) is part of an individual coping strategy of the animal and thus an important indicator of a trait-like behavioural and physiological response pattern. In their extensive studies on the endophenotypes of high-aggression and non-aggressive rodents, 5-HT was also studied. In contrast to the existing theory of inversed relationship between 5-HT activity and aggression, a positive correlation was found between the level of trait-like aggression (high or low) and basal CSF concentrations of 5-HT and 5-HIAA (Van der Vegt et al 2003). Moreover, levels of 5-HT and 5-HIAA after microdialysis in the frontal cortex did not differ between endophenotypes. Apparently, normal offensive aggression is positively related to serotonergic neuronal activity, whereas an inverse relationship probably exists between 5-HT activity and impulse-like violent aggression (Coccaro 1989).

Thus a general pattern emerges where trait and state aggression are probably differentially regulated by the 5-HT system (and also other systems) although much more research is needed to substantiate this hypothesis.

Serotonin receptors

The 14 different 5-HT receptors and the 5-HTT enable the unravelling of the function and contribution of the different receptors in various aspects of aggression (Table 1). To date, ligands are available for most receptors, including (partial) agonists and antagonists, although for some receptors no adequate tools are available, particularly for recently described receptors (5-$HT_{5,6,7,1e,1f}$). Since 5-HT_1 and $_2$ receptors have been studied since the 1980s, research on the role of these receptors in aggression is most abundant.

5-HT₁ receptors

Clinically, only one partial 5-HT_{1A} receptor agonist, buspirone has been available and its effect on aggression in humans is not very well described (Ratey et al 1991, Kavoussi et al 1997) but it does not appear very promising. Preclinically, an extensive range of 5-HT_{1A} receptor agonists and antagonists are available (Olivier et al 1999). And prototypic agonists like 8-OH-DPAT (8-hydroxy-2-di-n-propylamino-tetralin), flesinoxan and buspirone or antagonists like WAY100,635 have been used in aggression studies in animals.

Although practically all studies report anti-aggressive effects of 5-HT_{1A} receptor agonists in various species, these effects are not specific in the sense that anti-aggressive effects occur at doses that also compromise non-aggressive elements of the behavioural repertoire (Olivier et al 1995, Miczek et al 1998a,b). 5-HT_{1A} receptor antagonists have no intrinsic effects on aggression, but are able to antagonize the anti-aggressive effects of 5-HT_{1A} receptor agonists (Miczek et al 1998b, Mendoza et al 1999, De Boer et al 2000). There is conflicting evidence whether 5-HT_{1A} receptor agonists exert their anti-aggressive effects via pre- or postsynaptically located 5-HT_{1A} receptors. Lesions of the raphe nuclei, thereby removing 5-HT_{1A} somatodendritic autoreceptors, did not prevent the anti-aggressive effects of eltoprazine (Sijbesma et al 1991) or 8-OH-DPAT (Nikulina & Miczek 1999). However, eltoprazine is a mixed 5-$HT_{1A,1B}$ receptor agonist and its anti-aggressive effects in 5,7-DHT-lesioned rats might be due to activation of postsynaptic 5-HT_{1B} receptors. In the Nikulina & Miczek (1999) study, remaining 5-HT_{1A} autoreceptors (there was a limited depletion of serotonin in this study, suggesting a considerable number of intact 5-HT neurons after the 5,7-DHT insult) may be responsible for this effect. Infusion of 8-OH-DPAT and eltoprazine into the dorsal raphe nucleus (Mos et al 1993) reduced aggressive behaviour in rats but concomitantly reduced social interest and increased inactivity, indicative of a non-selective reduction of aggression. TFMPP (triflouromethylphenylpiperazine), a more selective 5-HT_{1B} receptor agonist than eltoprazine, had no effect under these conditions suggesting that the non-specific reduction of aggression after 8-OH-DPAT and eltoprazine was caused by activation of serotonergic autoreceptors in the dorsal raphe nucleus. When these drugs were infused into the lateral ventricle, 8-OH-DPAT had no anti-aggressive effects, whereas eltoprazine and TFMPP had a very selective anti-aggressive effect. This indicates that postsynaptic 5-HT_{1A} receptors are not involved in the aggression-modulating effects of 5-HT_{1A} receptor agonists and that the specific reduction in aggression is induced by activation of postsynaptically located 5-HT_{1B} receptors (Mos et al 1992). A strong argument against a specific role of 5-HT_{1A} receptors in specific modulation of (offensive) aggression derives from studies where aggression was evoked by electrical stimulation in the hypothalamus of rats (Kruk et al 1979). This kind of extremely aggressive behaviour towards a male conspecific was not at all sensitive to 5-HT_{1A} receptor agonists like 8-OH-DPAT, buspirone and flesinoxan (Olivier et al 1994), even at extremely high doses. Apparently, by directly stimulating the neural substrate activating aggressive behaviour, the role of the 5-HT_{1A} receptor is minimal or non-existing whereas 5-HT_{1B} receptor agonists (TFMPP and eltoprazine) dose-dependently and behaviourally specific, reduced this kind of 'violent' aggression.

On the other hand, certain 5-HT_{1A} receptor agonists from the benzodioxopiperazine class (Olivier et al 1999) seem to exert anti-aggressive effects via presynaptic 5-HT_{1A} autoreceptors (De Boer et al 2000, Van der Vegt et al 2001). Activation of presynaptic 5-HT_{1A} autoreceptors leads to decreased firing of the 5-HT neuron and decreased release of 5-HT at the synaptic level. It seems unlikely that such a general reduction in 5-HT turnover that affects all (pre- and postsynaptic) 5-HT receptors, would lead to such a specific reduction in aggression. Moreover, chronic administration of 5-HT_{1A} receptor agonists leads to down-regulation of 5-HT_{1A} autoreceptors and subsequent increased cell firing and enhanced 5-HT release. On a chronic base the acute anti-aggressive effects would subside and even enhanced aggression might develop after chronic treatment. No chronic studies with 5-HT_{1A} receptor agonists have been

TABLE 1 Summary of effects of psychoactive agents with selectivity for subtypes of serotonin receptors

5-HT principle	*Resident–intruder mouse*	*Resident–intruder rat*	*Maternal aggression rat*	*Brain-induced aggression rat*	*Muricide rat*	*Defensive behaviour mouse/rat*
1A agonist	0 ▼	▼	▼	0	0	0
1A antagonist	0	0	0	0	0	0
1B agonist	+	+	+	+	+	0
1B antagonist	0	0	0	0	0	0
2A agonist	▼	▼	▼	—	▼	—
2A antagonist	▼	▼	—	0	▼	—
2C agonist	▼	▼	▼	—	—	—
2C antagonist	▼	▼	▼	—	—	—
3-agonist	0	0	—	—	0	—
3-antagonist	0	0	0	—	0	—
Reuptake blockade	▼	▼	▼	0	0	0

0, no effect on aggression; ▼, non-specific reduction in aggression; +, specific reduction in aggression (serenic effect); —, not-tested.
Data are based on own published and unpublished work (summarized in Olivier et al 1990, 1994, 1995).

reported, but are clearly needed to answer the acute vs. chronic effects of 5-HT$_{1A}$ receptor agonists on aggression.

The 5-HT$_{1B}$ receptor has been the focus of interest in the development of clinically relevant anti-aggressive agents, the so-called serenics (Olivier et al 1990). The 5-HT$_{1B}$ receptor terminology has been confusing. The 5-HT$_{1B}$ receptor in rodents and humans, although exerting a similar function, differs in an essential amino acid in the ligand binding domain of the receptor that leads to a dramatic difference in the pharmacological sensitivity and specificity. Agonists for the rodent (r5-HT$_{1B}$) 5-HT$_{1B}$ receptor are anti-aggressive and exert a serenic profile (Olivier et al 1990), defined as a dose-dependent decrease in offensive aggression, without concomitant sedation, motor or sensory impairment that could explain the anti-aggressive effect. The early serenics (e.g. fluprazine, DU28412, DU 27725, eltoprazine, batoprazine; Olivier et al 1990) were mixed 5-HT$_{1A/1B}$ receptor agonist, leaving the 5-HT$_{1A}$ receptor still as an option for mediating (part of) the anti-aggressive effect, but more recently synthesized 5-HT$_{1B}$ receptor agonists including anpirtoline, CP-94,253 and zolmitriptan, were far more selective for this receptor and showed a similar, highly specific anti-aggressive effect, both in aggressive residential mice and in mice made more aggressive via low-doses of alcohol or social instigation (Fish et al 1999, De Almeida et al 2001, Miczek & De Almeida 2001).

5-HT$_{1B}$ receptor knockout mice (Saudou et al 1994) show enhanced aggressive behaviour (Saudou et al 1994, Ramboz et al 1995, Brunner & Hen 1997, Bouwknecht et al 2001), but due to the low baseline aggression level of the genetic background (129Sv) strain, the 'enhanced' aggression in the knockout mice was still low. More recent studies (Pattij et al 2003, 2004, Bouwknecht et al 2001) have implicated the 5-HT$_{1B}$ receptor in impulsivity regulation, rather than offensive aggression *per se* (Lesch & Merschendorf 2000); Olivier et al (1995) suggest that the specific anti-aggressive effects of 5-HT$_{1B}$ receptor agonist are modulated via postsynaptic 5-HT$_{1B}$ receptors. Such postsynaptic 5-HT$_{1B}$ receptors are located as heteroreceptors on non-serotonergic neurons (including dopaminergic, cholinergic, and GABA$_A$-ergic neurons). These heteroceptors, when activated by 5-HT inhibit ongoing behaviour, including aggression. Thus, 5-HT$_{1B}$ receptor agonist, inhibit those 'aggression or impulsivity' modulating neurons and removing the postsynaptic 5-HT$_{1B}$ receptor (via null mutation of the 5-HT$_{1B}$ receptor gene) removes this 'brake', thereby facilitating various behaviours related to impulsivity, hyperactivity and aggression (Olivier & Young 2002). Based on the 'hyperaggressive' 5-HT$_{1B}$ receptor knockout mouse and the anti-aggressive effects of 5-HT$_{1B}$ receptor agonists, one could suggest that administration of 5-HT$_{1B}$ receptor antagonists might lead to facilitation of aggression. However, all such antagonists appear silent, comparable to 5-HT$_{1A}$ receptor antagonists, probably indicating that under normal, physiological conditions the serotonergic tone at post-synaptic 5-HT_{1B} receptors is not that strong.

5-HT_2 receptors

Although 5-HT_2 ligands have been synthesized and tested in aggression, these compounds are extremely difficult to typify on their specificity in aggression (see for a nice review Miczek et al 2002). The available agents for 5-HT_{2A} and $_{2B}$ receptors are not that specific and when tested show anti-aggressive effects at similar doses that exert sometimes severe (sedating) side effects (Olivier et al 1995, Sanchez et al 1993). There is some evidence that 5-HT_{2A} receptor antagonists (e.g. risperidone) inhibit aggressive behaviour in patients, comprising various diagnostic areas, including depression and schizophrenia. However, no specific evidence thus far points to a selective contribution of any 5-HT_2 subtype receptor to aggression.

5-HT transporter

Since the 5-HTT is only located on the serotonergic neuron (both in the synaptic terminal and the somatodendritic areas), its function is directly related to the feedback mechanisms involved in 5-HT neuron firing and 5-HT release. Because specific serotonin re-uptake inhibitors (SSRIs) block the uptake of 5-HT, the net effect after acute SSRI administration is probably a mild increase in 5-HT release at the synaptic terminal. After chronic administration, leading to down-regulation of somatodendritic 5-HT_{1A} autoreceptors and hence lowered inhibition of cell-firing and thus enhanced terminal 5-HT release, 5-HT is probably far stronger enhanced than after acute administration. Therefore, acute, but certainly chronic administration of SSRIs should lead to inhibition of aggression, based on the hypothesis that activation of postsynaptic 5-HT_{1B} receptors mediates anti-aggressive effects. Clinical data seem to support this hypothesis (Coccaro & Kavoussi 1997, Hollander 1999, Cherek et al 2002). However, SSRIs are not known as mainstream treatment for aggressive pathology in humans. Preclinically, acute administration of SSRIs has anti-aggressive effects although not in a behaviourally specific way (Olivier et al 1995). This is not too surprising because the enhanced 5-HT activates all 5-HT receptors, and thus a plethora of effects might be expected. Chronic administration of SSRIs leads to contrasting effects: in mice (Delina-Stula & Vassoet 1978) to reduction, in rats however to increases (Mitchell & Redfern 1997). Mutant mice lacking the 5-HTT are less aggressive than their wild-types (Holmes et al 2003), confirming that chronically elevated 5-HT release is inhibitory on aggression. Although infrequent reports emerge on the possible role of other (5-$HT_{3,4,6,7}$) serotonergic receptors, it is unclear whether, and what role they might play.

Conclusions

The serotonergic system in the CNS has complex interactions with many, if not all other neurotransmitter systems in the brain. Its localization, distribution and amazing receptor diversity makes it an appealing system for modulatory aspects in many basic behaviours, including food and water intake, sexual behaviour and aggression. Notwithstanding decades of research into the putative role of the serotonergic system in aggression, no clear picture has emerged. It seems, dependent on state or trait, to be involved in either the performance or the termination of aggressive behaviours. Present technology appears not to be developed enough to give answers to these questions. Application of drugs, and particular selective ligands for certain subtype receptors, seems a more promising approach to unravelling the role of 5-HT in aggression. The (postsynaptic) 5-HT_{1B} and to a lesser extent, the 5-HT_{1A} receptor seems to play a prominent role, at least in rodents, in the modulation of (offensive) aggression.

References

Bonaventura P, Voorn P, Luyten WHML, Jurzak M, Schotte A, Leyten JE 1998 Detailed mapping of serotonin 5-HT_{1B} and 5-HT_{1D} receptor messenger RNA and ligand binding sites in guinea pig brain and trigeminal ganglion: clues for function. Neuroscience 82:469–484

Boschert U, Amara DA, Segu L, Hen R 1994 The mouse 5-hydroxytryptamine1B receptor is localized predominantly on axon terminals. Neuroscience 58:167–82

Bouwknecht JA, Hijzen TH, Van der Gugten J, Maes RAA, Hen R, Olivier B 2001 Absence of 5-HT_{1B} receptors is associated with impaired motor impulse control in male 5-HT_{1B} knockout mice. Biol Psychiatry 49:557–568

Brown GL, Goodwin FK, Ballenger JC, Goyer PF, Major LF 1979 Aggression in humans: correlates with cerebrospinal fluid amine metabolites. Psychiatry Res 1:131–139

Brunner D, Hen R 1997 Insights into the neurology of impulsive behavior from serotonin receptor knockout mice. Ann NY Acad Sci 836:81–105

Chen HT, Clark M, Goldman D 1992 Quantitative autoradiography of [^{3}H]-paroxetine binding sites in rat brain. J Pharmacol Toxicol Meth 27:209–216

Cherek DR, Lane SD, Pietras CJ, Steinberg JL 2002 Effects of chronic paroxetine administration on measures of aggressive and impulsive responses of adult males with a history of conduct disorder. Psychopharmacology 159:266–274

Coccaro EF 1989 Central serotonin and impulsive aggression. Br J Psychiatry 155:52–62

Coccaro EF, Kavoussi RJ 1997 Fluoxetine and impulsive aggressive behavior in personality-disordered subjects. Arch Gen Psychiatry 54:1081–1088

D'Amato RJ, Largent BL, Snowman AM, Snyder SH 1987 Selective labelling of serotonin uptake sites in rat brain by [3H]-citalopram contrasted to labelling of multiple sites by [3H]-imipramine. J Pharmacol Exp Ther 112:121–128

De Almeida RMM, Nikulina E, Faccidomo S, Fish E, Miczek KA 2001 Zolmitriptan, a 5-$HT_{1B/1D}$ receptor agonist, alcohol, and aggression in male mice. Psychopharmacology 157:131–141

De Boer S, Van der Vegt BJ, Koolhaas JM 2003 Individual variation in aggression of feral rodent strains: a standard for the genetics of aggression and violence. Behav Genet 33:485–501

De Boer SF, Lesourd M, Mocaer E, Koolhaas JM 2000 Somatodendritic 5-HT_{1A} autoreceptors mediate the anti-aggressive actions of 5-HT(1A) receptor agonists in rats: an ethopharmacological study with S-15535, alnespirone, and WAY-100635. Neuropsychopharmacology 23:20–33
Delina-Stula A, Vassoet A 1978 The effects of antidepressants on aggressiveness induced by social deprivation in mice. Pharmacol Biochem Behav 14(suppl 1):33–41
Ferrari PF, van Erp AMM, Tornatzky W, Miczek KA 2003 Accumbal dopamine and serotonin in anticipation of the next aggressive episode in rats. Eur J Neurosci 17:371–378
Fish EW, Faccidomo S, Miczek KA 1999 Aggression heightened by alcohol or social instigation in mice: reduction by the 5-HT_{1B} receptor agonist CP-94,253. Psychopharmacology 146: 391–399
Giacalone E, Tansella M, Valzelli L, Garattini S 1968 Brain serotonin metabolism in isolated aggressive mice. Biochem Pharmacol 17:1315–1327
Gothert M, Propping P, Bonisch H, Bruss M, Nothen MM 1998 Genetic variation in human 5-HT receptors: potential pathogenetic and pharmacological role. Ann N Y Acad Sci 861:26–30
Grailhe R, Bosschert U, Hen R 1997 The 5-HT_5, 5-HT_6 and 5-HT_7 receptors. In: Olivier B, van Wijngaarden I, Soudijn W (eds) Serotonin receptors and their ligands. Pharmacochemistry Library, Vol 27, Elsevier Science BV, Amsterdam, p 311–323
Grossman CJ, Kilpatrick GJ, Bunce KT 1993 Development of a radioligand binding assay for the 5-HT_4 receptor: use of a novel antagonist. Br J Pharmacol 108:618–624
Hartig PR, Hoyer D, Humphrey PP, Martin GR 1996 Alignment of receptor nomenclature with the human genome: classification of 5-HT_{1B} and 5-HT_{1D} receptor subtypes. Tr Pharmacol Sci 17:103–105
Hoffman B 1993 Molecular biology of serotonin uptake sites. In: Van Houtte PM, Saxena PR, Paoletti R, Brunello N, Jackson AS (eds) Serotonin — from cell biology to pharmacology and therapeutics. Kluwer Academic, Dordrecht, p 9–19
Hollander E 1999 Managing aggressive behavior in patients with obsessive-compulsive disorder and borderline personality disorder. J Clin Psychiatry 60(suppl 15):38–44
Holmes A, Murphy DL, Crawley JN 2003 Abnormal behavioral phenotypes of serotonin transporter knockout mice: parallels with human anxiety and depression. Biol Psychiatry 54:953–959
Hrdina PD, Foy B, Hepner A, Summers RJ 1990 Antidepressant binding sites in brain: autoradiographic comparison of [^{3}H]paroxetine and [^{3}H]imipramine localization and relationship to serotonin transporter. J Pharmacol Exp Ther 252:410–418
Jacobs B, Azmitia E 1992 Structure and function of the brain serotonin system. Physiol Rev 72:165–229
Kavoussi R, Armstead P, Coccaro E 1997 The neurobiology of impulsive aggression. Psychiatry Clin North Am 20:395–403
Kruesi MJP, Rapoport JL, Hamburger S et al 1990 Cerebrospinal fluid monoamine metabolites, aggression, and impulsivity in disruptive behavior disorders of children and adolescents. Arch Gen Psychiatry 47:419–426
Kruk MR, van der Poel AM, de Vos-Frerichs TP 1979 The induction of aggressive behaviour by electrical stimulation in the hypothalamus of male rats. Behaviour 70:292–322
Lesch KP, Merschdorf U 2000 Impulsivity, aggression, and serotonin: a molecular psychobiological perspective. Behav Sci Law 18:581–604
Linnoila M, Virkkunen M, Scheinin M, Nuutila A, Rimon R, Goodwin FK 1983 Low cerebrospinal fluid 5-hydroxyindole-acetic acid concentration differentiates impulsive from nonimpulsive violent behaviour. Life Sci 33:2609–2614
Mann JJ 2003 Neurobiology of suicidal behaviour. Nat Rev Neurosci 4:819–828
Mendoza DL, Bravo HA, Swanson HH 1999 Antiaggressive and anxiolytic effects of gepirone in mice, and their attenuation by WAY-100,635. Pharmacol Biochem Behav 62:499–509

Miczek KA, de Almeida RMMA 2001 Oral drug self-administration in the home cage of mice: alcohol-heightened aggression and inhibition by the 5-HT_{1B} agonist anpirtoline. Psychopharmacology 157:421–429
Miczek KA, Barros HM, Sakoda L, Weerts EM 1998a Alcohol and heightened aggression in individual mice. Alcohol Clin Exp Res 22:1698–1705
Miczek KA, Hussain S, Faccidomo S 1998b Alcohol-heightened aggression in mice: attenuation by 5-HT_{1A} receptor agonists. Psychopharmacology 139:160–168
Miczek KA, Fish EW, de Bold JF, de Almeida RMM 2002 Social and neural determinants of aggressive behavior: pharmacotherapeutic targets at serotonin, dopamine and γ-aminobutyric acid systems. Psychopharmacology 163:434–458
Mitchell PJ, Redfern PH 1997 Potentiation of the time-dependent, antidepressant-induced changes in the agonistic behaviour of resident rats by the 5-HT_{1A} receptor antagonist, WAY-100635. Behav Pharmacol 8:585–606
Mos J, Olivier B, Poth M, van Aken H 1992 The effects of intraventricular administration of eltoprazine, 1-(3-trifluoromethylphenyl) piperazine hydrochloride and 8-hydroxy-2-(di-n-propylamino) tetralin on resident intruder aggression in the rat. Eur J Pharmacol 212:295–298
Mos J, Olivier B, Poth M, van Oorschot R, Van Aken H 1993 The effects of dorsal raphe administration of eltoprazine, TFMPP and 8-OH-DPAT on resident intruder aggression in the rat. Eur J Pharmacol 238:411–415
Nikulina EM, Miczek KA 1999 Post- vs presynaptic sites of action of 5-HT_{1A} and 5-HT_{1B} receptor agonists in regulation of mouse aggressive behaviour. Behav Pharmacol 10: S67
Olivier B, Young L 2002 Animal models of aggression. In: Davis KL, Charney D, Coyle JT, Nemeroff C (eds) Neuropsychopharmacology: the 5th generation of progress. Lippincott, Williams & Wilkins, p 1699–1708
Olivier B, Mos J, Hartog J, Rasmussen DL 1990 Serenics: a new class of drugs for putative selective treatment of pathological destructive behaviour. Drug News Persp 3:261–271
Olivier B, Mos J, Raghoebar M, de Koning P, Mak M 1994 Serenics. Prog Drug Res 42: 167–308
Olivier B, Mos J, van Oorschot R, Hen R 1995 Serotonin receptors and animal models of aggressive behaviour. Pharmacopsychiatry 28:80–90
Olivier B, Soudijn W, van Wijngaarden I 1999 The 5-HT_{1A} receptor and its ligands: structure and function. Prog Drug Res 52:103–166
Palacios JM, Waeber C, Hoyer D, Mengod O 1990 Distribution of serotonin receptors. Ann NY Acad Sci 600:36–52
Pattij T, Van der Linde J, Groenink L et al 2003 Operant learning and differential-reinforcement-of-low-rate 36-sec responding in 5-HT_{1A} and 5-HT_{1B} receptor knockout mice. Behav Brain Res 141:137–145
Pattij T, Broersen L, Olivier B 2004 Impulsive-like responding in differential-reinforcement-of-low-rate 36-sec responding in mice depends on training history. Neurosci Lett 354:169–171
Piñeyro G, Blier P 1999 Autoregulation of serotonin neurons: role in antidepressant drug action. Pharmacol Rev 51:533–591
Plassat JL, Bosschert U, Amlaiky N, Hen R 1992 The mouse 5-HT_5 receptor reveals a remarkable heterogeneity within the 5-HT_{1D} receptor family. EMBO J 11:4779–4786
Ramboz S, Saudou F, Amara DA et al 1995 5-HT_{1B} receptor knock out-behavioral consequences. Behav Brain Res 73:305–312
Ratey J, Sovner R, Parks A, Rogentine K 1991 Buspirone treatment of aggression and anxiety in mentally retarded patients: a multiple-baseline, placebo lead-in study. J Clin Psychiatry 52:159–162
Sanchez C, Arnt J, Hyttel J, Moltzen EK 1993 The role of serotonergic mechanisms in inhibition of isolation-induced aggression in male mice. Psychopharmacology 110:53–59

Saudou F, Amara DA, Dierich A et al 1994 Enhanced aggressive behavior in mice lacking 5-HT_{1B} receptor. Science 265:1875–1878
Sijbesma H, Schipper J, de Kloet ER, Mos J, van Aken H, Olivier B 1991 Postsynaptic 5-HT_1 receptors and offensive aggression in rats: a combined behavioural and autoradiographic study with eltoprazine. Pharmacol Biochem Behav 38:447–458
Törk I 1990 Anatomy of the serotonergic system. Ann NY Acad Sci 90:9–35
Van der Vegt BJ, de Boer SF, Buwalda B, de Ruiter AJ, de Jong JG, Koolhaas JM 2001 Enhanced sensitivity of postsynaptic serotonin-1A receptors in rats and mice with high trait aggression. Physiol Behav 74:205–211
Van der Vegt BJ, Lieuwes N, Cremers TIFH, De Boer SF, Koolhaas JM 2003 Cerebrospinal fluid monoamine and metabolic concentrations and aggression in rats. Horm Behav 44:199–208
Van Erp MM, Miczek KA 2000 Aggressive behavior, increased accumbal dopamine and decreased cortical serotonin in rats. J Neurosci 15:9320–9325
Zifa E, Fillion G 1992 5-Hydroxytryptamine receptors. Pharmacol Rev 44:401–458

DISCUSSION

Martinez: Serenics reduced offensive but not defensive aggression. However, they also reduced maternal aggression. Is that correct? It seems surprising.

Olivier: Yes. You could say that the attacks of the female on the male were offensively motivated. The function may be defence, but the performance is clearly offensive behaviour. She sees an approaching animal and she attacks it, and this can be inhibited by 5-HT_{1B} agonists, although not that well by 5-HT_{1A} agonists. Some people have suggested in the past that if you treat an animal with eltoprazine or a serenic and it decreases a behaviour, then this behaviour must be offensive.

C Blanchard: Stefano Parmigiani has also done studies looking at the form of attack by females towards male and female intruders (Parmigiani et al 1989). The difference may have been that he was using adult males, whereas clearly the animal in your study was not an adult male.

Olivier: It was a male of approximately 200 g. The females are this sort of weight, and if you put a 500 g male in there it is not a fair fight.

C Blanchard: But that is why you get the defensive attack by the female. There appears to be a difference when you use a less threatening vs. a more threatening male intruder, based on both behaviour and on fluprazine effects. Parmigiani got this differentiation in maternal aggression. We got it in resident–intruder aggression: fluprazine reduced offensive but not defensive attack (Racine et al 1984).

Olivier: Eltoprazine and other serenics work on the offensive part of the system but not on the defensive part. This was our aim. We synthesized something like 10 000 compounds and tested them, and the result was these kinds of compounds which appear to be 5-HT_{1B} agonists.

Brodkin: I am a little puzzled by the pharmaceutical industry's argument that serenics should not be developed because aggression is a symptom, not a disease. Most psychiatric medications are not truly disease-specific.

Olivier: But they were developed for a specific disease, and then they are used for other things.

Brodkin: Clinically, psychiatrists worry a great deal about risks of impulsive aggression, e.g. risks of suicidality or homicidality. Concerns about impulsive aggression are the reason for many psychiatric hospitalizations and the source of enormous healthcare costs.

Ferris: How much of this impulsive aggression is due to mental illness, and how much is due to a social problem?

Brodkin: That's an important question, and difficult to answer precisely. Social problems can certainly contribute to aggressive behaviours. Mental illnesses, such as mood or psychotic disorders, also increase the risk of impulsive aggression directed against the self (suicidality) (Mann 2003). Whether mental illnesses increase risk of impulsive aggression directed towards others is an area of controversy, but some studies indicate that actively symptomatic psychosis and/or substance abuse, for example, can increase the risk of impulsive aggression towards others (see for example Taylor et al 1998, Steadman et al 1998).

Olivier: Aggression has frequently been considered as a positive attitude. People don't like the term anti-aggressive, which is why we invented the term 'serenic'.

C Blanchard: This is certainly part of American culture. If you acknowledge that the focal function of aggression involves resource competition, America was founded on aggression. The revolutionary war was a war to control the resources of America and by extension the rights of Americans. It could not have been successfully fought without an attitude that individuals have a right to bear arms and to use these when they feel it is necessary. This has had lots of implications for the culture and the laws of the USA.

Ferris: How far did you get in your clinical trials? What was the indication?

Olivier: We went to early phase 3. We did complete phase 1 and phase 2, and we did a lot of pilots in phase 2 studies with many different small populations of patients with all kinds of aggressive troubles. We wanted to first develop scales: this was part of the problem. There were no specialists investigating violent aggression. Finally, we decided to go into severely aggressive mentally retarded patients, with a lot of self-injurious behaviour. They weren't on drugs. We had 150 patients in a double-blind placebo-controlled cross-over trial. It was effective, but we had a severe placebo response which we had not expected. You would not expect that these kinds of patients would have insight into their treatment. Afterwards we realized that the expectations of the nurses, care-givers and parents were very high. These people got a lot of attention, and for these kinds of patients attention is the best serenic.

Manuck: Perhaps an appropriate clinical entity for testing serenics would be Intermittent Explosive Disorder, in which a patient's inability to resist aggressive impulses occasions episodes of physical assault or property destruction that are inexplicable as responses to provocation.

With respect to genetics, is there any known functional variation in the gene encoding the 5-HT_{1B} receptor?

Lesch: There are some coding variants which are very rare. There are also some anonymous single nucleotide polymorphisms (SNPs) in the coding region. A recent publication (Duan et al 2003) describes an interesting haplotype in the promoter region of the 5-HT_{1B} receptor gene (*HTR1B*) which is functional and could be used for association studies.

Olivier: There is also a publication where they didn't find any difference.

Lesch: I think there was a positive finding in alcohol-dependent patients and antisocial behaviour (Lappalainen et al 1998).

Manuck: I know of four papers examining the 5-HT_{1B} receptor. A silent guanine-cytosine substitution at nucleotide 861 was first found in linkage with antisocial alcoholism in two populations: alcoholic criminal offenders and their siblings, in Finland, and a multigenerational family of Native Americans in the American Southwest (Lappalainen et al 1998). However, allelic association with antisocial alcoholism was not observed subsequently among Americans of either European or African heritage (Krantzler et al 2002). And in two other studies, this polymorphism was found unrelated to 'pathological aggression' (Huang et al 1999) and, among personality disordered patients, to dispositional hostility assessed by questionnaire (New et al 2001).

Olivier: I don't think necessarily something should be along the 5-HT_{1B} receptor. It could be in the neurons themselves. Finally, what matters is the amount of serotonin that arrives at the postsynaptic 5-HT_{1B} receptor. This can be enhanced or lowered. Like your differing lengths of the transporter promoter, this could influence the amount of serotonin available for the postsynaptic receptor, and this could explain how people respond under certain conditions.

Keverne: In your studies the animals showed quite a bit of time investigating before they made any attack. It wasn't impulsive or spontaneous aggression. You also showed that these drugs affect sexual behaviour. If you use a 5-HT_{1B} antagonist, sexual behaviour goes up, and the agonist reduces this. To what extent is the aggressive behaviour you see secondary to a change in the animals' sexual motivation?

Olivier: That is a good question. I don't believe that there is a correlation between sex and aggression in that sense.

Keverne: Aggressive behaviour that is not spontaneous is usually secondary to getting access to females or territories. It is not a primary behaviour.

Olivier: Maternal aggression is not driven by sexual behaviour. 5-HT_{1A} agonists reduce aggressive behaviour and stimulate sexual behaviour. I would not expect that if they were similarly motivated.

Keverne: I also noticed that these drugs had effects on alcohol and cocaine: the agonist reduced them. This suggests that they are probably acting in the ventral striatum or nucleus accumbens.

Olivier: If we are discussing where in the brain this effect takes place, I would go for the basal ganglia. In the studies there are two males, the alpha male and the subordinate male. The alpha male normally does the attacks, and the other male joins in. If you treat the subordinate with low doses of alcohol or benzodiazepines, he takes over the role of the alpha male. The alpha male is not influenced by these doses. There is some anxiety or fear around in this situation.

Pfaff: I have a related question which can be stated in a more general form. Regarding the specificity of the actions of your drugs, obviously having considerable specificity is important for therapy. But, would you say that the several non-specific effects you showed reveal something scientifically about some of the mechanisms that actually underlie aggression to begin with?

Olivier: I think if you give an SSRI and it has a mild anti-aggressive effects along with all kinds of other effects on behaviour, this has to do with the activation of many more receptors than just the 5-HT_{1B} receptor. Serotonin is enhanced all over the brain by this SSRI, and it stimulates all the available receptors. Depending on the situation you are in I think you can see side effects. Perhaps they are not side effects but real effects in this situation.

Pfaff: I'll enter the alternative hypothesis: some of the mechanisms are on generalized arousal pathways.

Olivier: They could be.

Koolhaas: We are somehow challenging the serotonin deficiency hypothesis. We have a slightly different view. The hypothesis holds true if you look at serotonin as a trait characteristic. If we measure the strength of the negative feedback on the 5-HT_{1A} receptor, it has a 10-fold higher sensitivity in high aggressive males compared to the low aggressive males. We have demonstrated this in mice and rats. This means that as a trait characteristic the 5-HT neuron is strongly inhibited by autoreceptor control in the former group, which is consistent with the serotonin deficiency hypothesis. However, at the same time we have evidence that during the aggressive act itself, this strongly inhibited neuron gets released from its brakes and releases a large amount of 5-HT. The evidence we have comes from behavioural pharmacological data, and measuring raphe activity using c-Fos expression in the rat in combination with immunocytochemical staining of 5-HT. This latter approach shows that the act of aggressive behaviours is associated with an increase in activity of serotonergic cells in the raphe. We are currently trying to find further evidence that the serotonin deficiency hypothesis does not hold when it comes to the aggressive act itself.

Olivier: The neurotransmitter serotonin is present. It works somewhere else in the brain, not on the serotonergic cell itself. The serotonergic cell doesn't do anything other than release serotonin for postsynaptic receptors. What you are doing is preparing the serotonergic system for this postsynaptic 5-HT_{1B} receptor: this is my hypothesis.

Koolhaas: I agree with that. It still holds for the serotonin deficiency hypothesis.

Olivier: It is a fairly specialized serotonin deficiency hypothesis.

Koolhaas: It is fair to say we have this debate on somewhat conflicting data.

Suomi: In the situation where the alpha and beta male exhibit differential behaviour toward the intruder, have you tried to take the alpha male out of the game by sedating it or physically removing him?

Olivier: Yes, and then the beta immediately takes over.

R Blanchard: To what extent do the fluvoxamine results match the results obtained with the benzodiazepines? Is it the relief of the inhibition of aggression by defensive modulation, rather than a primary effect? Low doses of valium potentiate aggression.

Olivier: If there are two males in a cage, the beta males stay sensitive to alcohol and benzodiazepines, but not towards fluvoxamine.

Brodkin: I know that many of these compounds aren't available currently. But for those that are available for clinical use, lithium and clozapine are among the most effective agents for preventing impulsive aggression (Fava 1997, Mann 2003, Volavka et al 2004). How do these fit into a 5-HT_{1B} mechanism?

Olivier: Lithium is a fairly general drug that could work behind receptors on some mechanisms behind serotonin. It is a dirty drug that is difficult to handle, so I wouldn't treat aggressive patients with it.

Ferris: It is routinely dispensed to prison inmates with a history of impulsivity and violence.

Olivier: Clozapine is a fairly sedative compound. Treating people with clozapine tends to sedate people lowering the risk that they will do some active acts, rather than acting specifically.

R Blanchard: How would you approach the problem today differently than you did 10 years ago?

Olivier: First of all, I would try to gather a good group of psychiatrists who deal with this kind of work. There are more people around now than in those days. Then I would try to get good populations of intermittent explosive disorder. This wasn't well defined in the 1980s. This hindered our research.

Ferris: I am familiar with a drug company that has funding from the NIH to develop serenics. They have confined the indication to autism and tourettes, and the aggression that is associated with these primary conditions.

Olivier: That is an orphan drug development, supported by the government.

Ferris: This is one way of getting resources to do this kind of needed science.

Skuse: There have been some rather good trials on the use of Risperidone (McCracken et al 2002) which is a clozapine-like drug, with a major impact as an agonist on 5-HT_{1A} receptors (Meltzer et al 2003). It is very effective at reducing aggression in conditions such as autism without any sedation.

Olivier: It has never been developed for that.

Skuse: No, but it has been widely used clinically because it is so effective. It targets the aggression and appears not to have any other impact on behavioural cognition.

Olivier: Basically it is a dopamine D2 blocker. You could also use Haloperidol.

Skuse: Haloperidol tends to be sedative, and is associated with the development of dyskinesias (Posey & McDougle 2000). We would use it if it was found to be as effective as Risperidone.

References

Duan J, Sanders AR, Molen JE et al 2003 Polymorphisms in the 5′-untranslated region of the human serotonin receptor 1B (HTR1B) gene affect gene expression. Mol Psychiatry 8:901–910

Fava M 1997 Psychopharmacologic treatment of pathologic aggression. Psychiatric Clin North Amer 20:427–451

Huang Y, Grailhe R, Arango V, Hen R, Mann JJ 1999 Relationship of psychopathology to the human serotonin1B genotype and receptor biding kinetics in post-mortem brain tissues. Neuropsychopharmacology 21:238–246

Krantzler HR, Hernandez-Avila CA, Gelernter J 2002 Polymorphism of the 5-HT1B receptor gene (HTR1B): Strong within-locus linkage disequlilbrium without association to antisocial substance dependence. Neuropsychopharmacology 26:115–122

Lappalainen J, Long JC, Eggert M et al 1998 Linkage of antisocial alcoholism to the serotonin 5-HT1B receptor gene in 2 populations. Arch Gen Psychiatry 55:989–994

Mann JJ 2003 Neurobiology of suicidal behaviour. Nat Rev Neurosci 4:819–828

McCracken JT, McGough J, Shah B et al 2002 Risperidone in children with autism and serious behavioral problems. N Engl J Med 347:314–321

Meltzer HY, Li Z, Kaneda Y, Ichikawa J 2003 Serotonin receptors: their key role in drugs to treat schizophrenia. Prog Neuropsychopharmacol Biol Psychiatry. 27:1159–1172

New AS, Gelernter J, Goodman M et al 2001 Suicide, impulsive aggression, and HTR1B genotype. Biol Psychiatry 50:62–65

Parmigiani S, Rodgers RJ, Palanza P, Mainardi M, Brain PF 1989 The inhibitory effects of fluprazine on parental aggression in female mice are dependent upon intruder sex. Physiol Behav 46:455–459

Posey DJ, McDougle CJ 2000 The pharmacotherapy of target symptoms associated with autistic disorder and other pervasive developmental disorders. Harv Rev Psychiatry 8:45–63

Racine MA, Flannelly KJ, Blanchard DC 1984 Anti-aggressive effects of DU 27716 on attack and play fighting of rats in comparative perspective: a schema for neurobehavioral analyses. Prog Clin Biol Res 169:281–293

Steadman HJ, Mulvey EP, Monahan J et al 1998 Violence by people discharged from acute psychiatric inpatient facilities and by others in the same neighbourhoods. Arch Gen Psychiatry 55:393–401

Taylor PJ, Leese M, Williams D, Butwell M, Daly R, Larkin E 1998 Mental disorder and violence: a special (high security) hospital study. Brit J Psychiatry 172:218–226

Volavka J, Czobor P, Nolan K et al 2004 Overt aggression and psychotic symptoms in patients with schizophrenia treated with clozapine, olanzapine, risperidone, or haloperidol. J Clin Psychopharmacol 24:225–228

Vasopressin/oxytocin and aggression

Craig F. Ferris

Center for Comparative Neuroimaging, University of Massachusetts Medical School, Medical School 55 Lake Avenue North, Worcester, MA 01655, USA

Abstract. Vasopressin/oxytocin and related peptides comprise a phylogenetically old superfamily of chemical signals in both vertebrates and invertebrates. Each peptide isoform has its own distinct receptor subtype and specific cellular action. The conservation and dispersion of vasopressin/oxytocin signalling systems across the animal kingdom attests to their functional significance in evolution. Indeed, they are involved in the physiology of fluid balance, carbohydrate metabolism, thermoregulation, immunity and reproduction. In addition, these peptides evolved a role in social behaviours related to aggression and affiliation. The focus of this chapter is the role of vasopressin/oxytocin as chemical signals in the brain altering aggressive responding in a context- and species-dependent manner. There is compelling evidence from several mammalian species including humans that vasopressin enhances aggression. The activity of the vasopressin appears linked to the serotonin system providing a mechanism for enhancing and suppressing aggressive behaviour.

2005 Molecular mechanisms influencing aggressive behaviours. Wiley, Chichester (Novartis Foundation Symposium 268) p 190–200

Neurochemical control of aggression

Oxytocin

Oxytocin (OT) is heterogeneously dispersed throughout the brain with the limbic system and ventral striatum having a particularly high concentration and expression of OT receptor. The pattern of OT receptor binding differs considerable across species and may be predictive of behavioural phenotypes (Insel & Shapiro 1994). For example, prairie voles (*Microtus ochrogaster*) form long-term monogamous relations following mating. In addition to this pair bonding preference, male prairie voles become highly aggressive toward other male intruders. This monogamous, aggressive behaviour may be controlled, in part, by OT since the prairie vole and other *Microtus* species with similar behavioural phenotypes show a preferential distribution of OT in the prelimbic cortex, bed nucleus of the stria terminalis, nucleus accumbens, and the lateral aspects of the amygdala. These brain areas showed little OT binding in other *Microtus* species such as the montane vole (*Microtus montanus*) that are polygamous and less aggressive. Consequently, these diverse behavioural

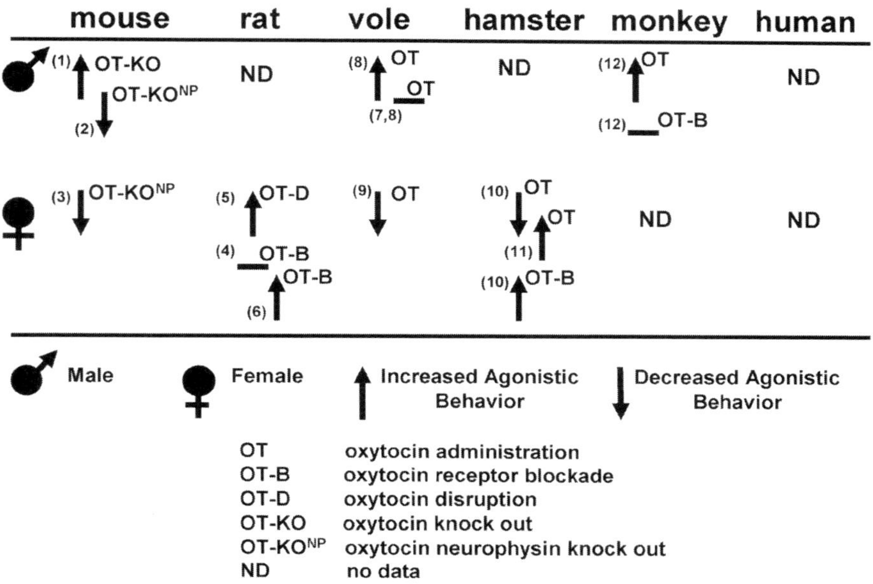

FIG. 1. Oxytocin and agonistic behaviour. References cited in figure: (1) Winslow et al 2000 (Horm Behav 37:145–155), (2) DeVries et al 1997 (J Neuroendocrinol 9:363–368), (3) Young et al 1998 (Adv Exp Med Biol 449:231–240), (4) Neumann et al 2001 (Eur J Neurosci 13:1016–1024), (5) Giovenardi et al 1998 (Physiol Behav 63:351–359), (6) Lubin et al 2003 (Behav Neurosci 117:195–201), (7) Mahalati et al 1991 (Pharmacol Biochem Behav 39:219–222), (8) Winslow et al 1993 (Psychopharmacol Bull 29:409–414), (9) Witt et al 1990 (Pharmacol Biochem Behav 37:63–69), (10) Harmon et al 2002 (J Neuroendocrinol 14:963–996), (11) Ferris et al 1992, (12) Winslow & Insel 1991 (J Neurosci 11:2032–2038).

phenotypes respond differently to manipulations that alter oxytocinergic neurotransmission. Virgin male prairie and montane voles show little if any change in aggressive behaviour to intracerebroventricular (ICV) administration of OT (Winslow et al 1993a). However, following mating, prairie voles become very aggressive in response to OT administration while montane voles remain unresponsive. The results shown that species differences and social experience alters OT's behavioural effects.

Figure 1 refers to different studies reporting manipulations in oxytocinergic neurotransmission and changes in aggression across multiple species. There appears to be no predictable response as OT promotes aggression in some cases and inhibits aggression in others. For example, resident, adult male OT knockout mice show more aggressive behaviour than wild-type controls toward male intruders (Winslow et al 2000). This finding would suggest that oxytocin in wild-types contributes to a reduction in offensive aggression. However, targeted elimination of the mouse OT neurophysin gene causing a deficiency in OT is reported to reduce aggression in the resident-intruder paradigm (DeVries et al 1997). One explanation for these disparate results is the reduction in brain levels of VP following deletion of the OT-associated neurophysin. In early postpartum dams, lesioning OT neurons in the paraventricular nucleus or reducing OT synthesis by antisense injection increases maternal aggressive behaviour (Giovenardi et al 1998). However, these effects are dependent on the reproductive condition of the female because similar manipulations during late postpartum, as pups start to wean, have little effect on maternal aggression. Blockade of OT receptors in the central nucleus of the amygdala increases maternal aggression in rats (Lubin et al 2003) suggesting endogenous neuropeptide inhibits aggressive responding. However, OT injection in this nucleus of hamsters increases maternal aggression (Ferris et al 1992). Blockade of OT receptors in the medial preoptic area/anterior hypothalamus increases aggression in resident, female hamsters toward intruders while OT injection in the same area reduces aggression (Harmon et al 2002). Work on pair-housed squirrel monkeys report ICV OT can increase the frequency of aggressive displays in the dominant member without altering the behaviour of the subordinate (Winslow & Insel 1991). Pretreatment with OT receptor antagonist blocks the effect of ICV OT but given alone has no effect on aggressive behaviour suggesting endogenous OT is not affecting normal agonistic behaviour in this social dyad. Collectively, these studies do not suggest a common behavioural effect for OT and aggressive responding. Instead, OT's effect is highly variable and dependent upon species, gender, reproductive condition and environmental context.

Vasopressin

Vasopressin (VP) like OT, has a heterogeneous distribution in the brain with a particularly high density of fibres and receptors in limbic areas. The behavioural effects of VP are mediated through V1a or V1b receptors. Unlike OT, there appears to be a general consensus across species that VP can act at multiple brain areas to facilitate aggression (Fig. 2). Microinjection of VP into the anterior or ventrolateral hypothalamus of resident hamsters significantly increases the number of biting attacks on intruders (Delville et al 1995, Ferris et al 1997, Caldwell & Albers 2004). Infusion of VP into the amygdala or lateral septum facilitates attack behaviour in castrated rats (Koolhaas et al 1990, 1991). Prairie voles show a dose dependent increase in aggression toward intruders following ICV administration of VP (Young et al 1997). Data from rats and humans show high indexes of aggressivity correlate with high concentrations of VP in cerebrospinal fluid (Haller et al 1996, Cocarro et al 1998). In hamsters, a V1a

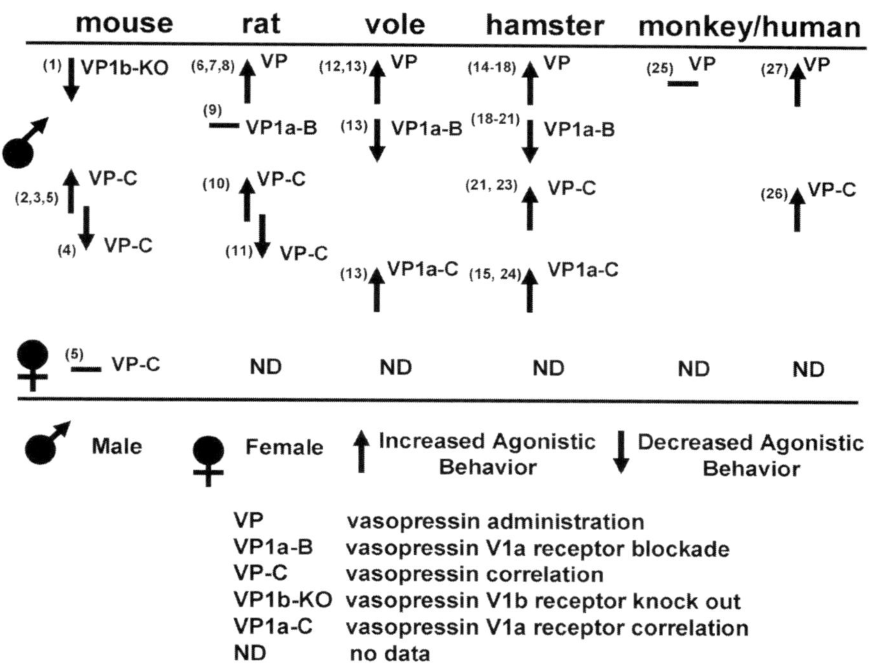

FIG. 2. Vasopressin and agonistic behaviour. References cited in figure: (1) Wersinger et al 2002 (Mol Psychiatry 7:975–984), (2) Bester-Meredith et al 1999 (Horm Behav 36:25–38), (3) Bester-Meredith & Marler 2001 (Horm Behav 40:51–64), (4) Compaan et al 1993 (Brain Res Bull 30:1–6), (5) Scordalakes & Rissman 2004 (Genes Brain Behav 3:20–26), (6) Elkabir et al 1990 (Regul Pept 28:199–214), (7) Koolhaas et al 1990 (Aggress Behav 16:223–229), (8) Koolhaas et al 1991 (Vasopressin. John Libbey, p 213–219) (9) Koolhaas et al 1998 (Prog Brain Res 119:437–448), (10) Haller et al 1996 (J Neuroendocrinol 8:361–365), (11) Everts et al 1997 (Horm Behav 31:136–144), (12) Winslow et al (Nature 365:545–548), (13) Young et al 1997 (Behav Neurosci 111:599–605), (14) Delville et al 1995 (Physiol Behav 59:813–816), (15) Delville et al 1996 (Physiol Behav 60:25–29), (16) Ferris et al 1997 (J Neurosci 17:4331–4340), (17) Ferris et al 1999 (Behav Neurosci 113:804–815), (18) Caldwell & Albers 2004 (Horm Behav 46:444–449), (19) Ferris & Potegal 1988 (Physiol Behav 44:235–239), (20) Potegal & Ferris 1990 (Aggress Behav 15:311–320), (21) Harrison et al 2000 (Psychoneuroendocrinol 25:317–338), (22) Ferris et al 1989 (Neurosci 29:675–683), (23) Delville et al 1998 (J Neurosci 18:2667–2672, (24) DeLeon et al 2002 (Horm Behav 42:182–191), (25) Winslow & Insel 1991 (Eur J Pharmacol 200:95–101), (26) Coccaro et al 1996 (Arch Gen Psychiatry 55:708–714), (27) Thompson et al 2004 (Psychoneuroendocrinol 29:35–48).

receptor antagonist, microinjected into the anterior hypothalamus, causes a dose-dependent inhibition of aggression of a resident male toward an intruder (Ferris & Potegal 1988). Treatment with V1a receptor antagonist prolongs the latency to bite an intruder, reduces the number of bites, but does not alter other social or appetitive behaviours. Vasopressin V1a receptor antagonist also blocks aggression associated with the development of dominant/subordinate relationships (Potegal & Ferris 1990). In male prairie voles, ICV injection of V1a receptor antagonist reduces aggression toward male intruders (Winslow et al 1993b). While all studies to date have focused on the blockade of V1a receptor in the regulation of aggression, evidence from knockout mice suggests that the V1b receptor may also be involved in facilitating aggressive behaviour (Wersinger et al 2002).

The VP system and the aggressive promoting effects of this neuropeptide are affected by photoperiod, early endocrine manipulations and social experience. Golden hamsters maintained on long photoperiods show high levels of sexual activity and VP regulated control of intermale aggression (Caldwell & Albers 2004). Injections of VP and V1a receptor antagonist into the anterior hypothalamus enhance and suppresses; respectively, attack behaviour toward intruders. However, this VP-mediated regulation of aggression is uncoupled during short photoperiods. With short photoperiods, golden hamsters show reduced sexual activity concomitant with lower levels of testosterone but exaggerated intermale aggression. The enhanced aggression under this environmental condition is not affected by VP or V1a blockade (Caldwell & Albers 2004). Thus, intermale aggression associated with high testosterone levels and competition for females may utilize a VP-sensitive neural pathway. During non-reproductive periods and reduced gonadal steroids, the neural pathway(s) responsible for aggression involving competition for resources do not require VP neurotransmission.

Testosterone treatment in castrated hamsters increases V1a receptor density in the ventromedial hypothalamus and VP-induced intermale aggression (Delville et al 1996). Treating adolescent hamsters with anabolic steroids increases the density of VP immunoreactive fibres, V1a receptors and neuropeptide content in the anterior hypothalamus and enhances VP mediated aggression as adults (Harrison et al 2000, DeLeon et al 2002). Conversely, social subjugation of adolescent hamsters reduces VP immunoreactivity in the anterior hypothalamus resulting in inappropriate aggressive behaviour. Hamsters subjugated in adolescence are highly aggressive toward non-threatening small intruders but submissive towards equal sized intruders (Delville et al 1998). The submissive hamsters in adult dominant/subordinate dyads also have reduced levels of VP in the anterior hypothalamus (Ferris et al 1989).

The strong correlation in golden hamsters (*Mesocricetus auratus*) between VP immunoreactivity in the brain and aggressive responding also extends to some species of mice, but not others. For example, inbred strains of mice bred for short and long attack latencies show a negative correlation between VP fibre immunostaining in the septum and aggressivity (Compaan et al 1993). In contrast, the highly aggressive California mouse (*Peromyscus californicus*) has greater VP staining in the bed nucleus of the stria terminalis and VP receptor density in the lateral septum as compared to the less aggressive white-footed mouse (*Peromyscus leucopus*) (Bester-Meredith et al 1999). Moreover, when California mice are cross-fostered with white-footed parents they show a reduction in aggression in a resident–intruder paradigm and lower levels of VP in the bed nucleus as compared to their unfostered siblings (Bester-Meredith & Marler 2001). These studies show that environmental conditions associated with pup rearing can affect VP neurotransmission and behaviour.

Vasopressin/serotonin interactions

Defining the mechanisms and anatomical substrates underlying interactions between functionally opposed neurotransmitter systems is critical for understanding aggressive behaviour. There are many studies across a wide range of animals showing serotonin (5-HT) reduces aggression (see Olivier 2005, this volume). There is evidence that VP and 5-HT interact to affect aggressive responding. The anterior hypothalamus, the primary site of VP regulation of aggression, has a high density of 5-HT binding sites and receives a dense innervation of 5-HT fibres and terminals (Ferris et al 1997). The VP neurons in the anterior hypothalamus implicated in the control of aggression appear to be preferentially innervated by 5-HT (Ferris et al 1997, 1999). Intraperitoneal injection of fluoxetine blocks aggression facilitated by the microinjection of VP in the hypothalamus (Delville et al 1995, Ferris et al 1997). Fluoxetine elevates 5-HT and reduces VP levels in hypothalamic tissue in rats (Altemus et al 1992). Kia and coworkers (1996) reported intense immunocytochemical staining for 5-HT_{1A} receptors in the VP system of rats supporting the notion that activation of 5-HT_{1A} receptors can influence the activity of VP neurons. However, data suggest 5-HT can also block the activity of VP following its release in the hypothalamus as evidenced by the dose-dependent diminution of aggression with injections combining VP and 5-HT_{1A} receptor agonist. Enhanced aggression caused by activation of AVP V_{1A} receptors in the hypothalamus is suppressed by the simultaneous activation of 5-HT_{1A} receptors in the same site. It is not clear whether a common neuronal phenotype in the hypothalamus shares both receptor subtypes, or VP and 5-HT act on separate neurons in the hypothalamus.

These animal studies examining the interaction between VP and 5-HT are particularly relevant since Cocarro and coworkers (1998) reported a similar reciprocal relationship in human studies. Personality disordered subjects with a history of fighting and assault show a negative correlation for prolactin release in response to D-fenfluramine challenge, indication of a hyposensitive 5-HT system. Moreover, these same subjects show a positive correlation between CSF levels of VP and aggression. Thus, in humans, a hyposensitive 5-HT system may result in enhanced CNS levels of VP and the facilitation of aggressive behaviour.

Summary

The ability of VP to affect aggression at multiple sites in the CNS and in various mammalian species is evidence that this neurochemical system may have a broad physiological role enhancing arousal and attack behaviour during agonistic interactions. Serotonin and VP appear to play significant roles in the regulation of impulsivity and aggression. Serotonin reduces aggressive responding while VP enhances arousal and aggression in a context-dependent manner. There is compelling neuroanatomical, pharmacological and molecular data supporting an interaction between 5-HT and VP in the control of aggression. VP promotes aggression by enhancing the activity of the neural network controlling agonistic behaviour that is normally restrained by 5-HT.

Acknowledgements

These experiments were supported by grants MH 52280 from the NIMH. The contents of this review are solely the responsibility of the authors and do not necessarily represent the official views of the NIMH.

References

Altemus M, Cizza G, Gold PW 1992 Chronic fluoxetine treatment reduces hypothalamic vasopressin secretion in vitro. Brain Res 593:311–313

Bester-Meredith JK, Marker CA 2001 Vasopressin and aggression in cross-fostered California mice (*Peromyscus californicus*) and white-footed mice (*Peromyscus leucopus*). Horm Behav 40:51–64

Bester-Meredith JK, Young LJ, Marker CA 1999 Species differences in paternal behavior and aggression in peromyscus and their associations with vasopressin immunoreactivity and receptors. Horm Behav 36:25–38

Caldwell HK, Albers HE 2004 Effect of photoperiod on vasopressin-induced aggression in Syrian hamsters. Horm Behav 46:444–449

Compaan JC, Buijs RM, Pool CW, De Ruiter AJ, Koolhaas JM. 1993 Differential lateral septal vasopressin innervation in aggressive and nonaggressive male mice. Brain Res Bull 30:1–6

Coccaro EF, Kavoussi RJ, Hauger RL, Cooper TB, Ferris CF 1998 Cerebrospinal fluid vasopressin levels correlates with aggression and serotonin function in personality-disordered subjects. Arch Gen Psychiatry 55:708–714

DeLeon KR, Grimes JM, Melloni RH Jr 2002 Repeated anabolic-androgenic steroid treatment during adolescence increases vasopressin V(1A) receptor binding in Syrian hamsters: correlation with offensive aggression. Horm Behav 42:182–191

Delville Y, Mansour KM, Ferris CF 1995 Serotonin blocks vasopressin-facilitated offensive aggression: interactions within the ventrolateral hypothalamus of golden hamsters. Physiol Behav 59:813–816

Delville Y, Mansour KM, Ferris CF 1996 Testosterone facilitates aggression by modulating vasopressin receptors in the hypothalamus. Physiol Behav 60:25–29

Delville Y, Melloni RH Jr, Ferris CF 1998 Behavioral and neurobiological consequences of social subjugation during puberty in golden hamsters. J Neurosci 18:2667–2672

DeVries AC, Young WS 3rd, Nelson RJ 1997 Reduced aggressive behaviour in mice with targeted disruption of the oxytocin gene. J Neuroendocrinol 9:363–368

Ferris CF, Potegal M 1988 Vasopressin receptor blockade in the anterior hypothalamus suppresses aggression in hamsters. Physiol Behav 44:235–239

Ferris CF, Axelson JF, Martin AM, Roberge LR 1989 Vasopressin immunoreactivity in the anterior hypothalamus is altered during the establishment of dominant/subordinate relationships between hamsters. Neurosci 29:675–683

Ferris CF, Foote KB, Meltser HM, Plenby MG, Smith KL, Insel T 1992 Oxytocin in the amygdala increases maternal aggression In: Pederson CA, Caldwell J (eds) Oxytocin in maternal, sexual and social behavior. NY Acad Sci 652:468–469

Ferris CF, Melloni RH Jr, Koppel G, Perry KW, Fuller RW, Delville Y 1997 Vasopressin/serotonin interactions in the anterior hypothalamus control aggressive behavior in golden hamsters. J Neurosci 17:4331–4340

Ferris CF, Stolberg T, Delville Y 1999 Serotonin regulation of aggressive behavior in male golden hamsters (*Mesocricetus auratus*). Behav Neurosci 113:804–815

Giovenardi M, Padoin MJ, Cadore LP, Lucion AB 1998 Hypothalamic paraventricular nucleus modulates maternal aggression in rats: effects of ibotenic acid lesion and oxytocin antisense. Physiol Behav 63:351–359

Haller J, Makara GB, Barna I, Kovacs K, Nagy J, Vecsernyes M 1996 Compression of the pituitary stalk elicits chronic increases in CSF vasopressin, oxytocin as well as in social investigation and aggressiveness. J Neuroendocrinol 8:361–365

Harmon AC, Huhman KL, Moore TO, Albers HE 2002 Oxytocin inhibits aggression in female Syrian hamsters. J Neuroendocrinol 14:963–969

Harrison RJ, Connor DF, Nowak C, Nash K, Melloni RH Jr 2000 Chronic anabolic-androgenic steroid treatment during adolescence increases anterior hypothalamic vasopressin and aggression in intact hamsters. Psychoneuroendocrinol 25:317–338

Insel TR, Shapiro LE 1992 Oxytocin receptor distribution reflects social organization in monogamous and polygamous voles. Proc Natl Acad Sci USA 89:5981–5

Kia HK, Miquel M-C, Brisorgueil M-J et al 1996 Immunocytochemical localization of serotonin 1A receptors in the rat central nervous system. J Comp Neurol 365:289–305

Koolhaas JM, Van den Brink THC, Roozendal B, Boorsma F 1990 Medial amygdala and aggressive behavior: interaction between testosterone and vasopressin. Aggress Behav 16:223–229

Koolhaas JM, Moor E, Hiemstra Y, Bohus B 1991 The testosterone-dependent vasopressinergic neurons in the medial amygdala and lateral septum: involvement in social behaviour of male rats. In: Jard S, Jamison R (eds)Vasopressin. INSERM/John Libbey Eurotext Ltds, Londres, p 213–219

Lubin DA, Elliott JC, Black MC, Johns JM 2003 An oxytocin antagonist infused into the central nucleus of the amygdala increases maternal aggressive behavior. Behav Neurosci 117:195–201

Olivier B 2005 Serotonergic mechanisms in aggression. In: Molecular mechanisms influencing aggressive behaviour (Novartis Found Symp 268). Wiley, Chichester, p 171–189

Potegal M, Ferris CF 1990 Intraspecific aggression in male hamsters is inhibited by vasopressin receptor antagonists. Aggress Behav 15:311–320

Wersinger SR, Ginns EI, O'Carroll AM, Lolait SJ, Young WS 3rd 2002 Vasopressin V1b receptor knockout reduces aggressive behavior in male mice. Mol Psychiatry 7:975–984

Winslow JT, Insel TR 1991 Social status in pairs of male squirrel monkeys determines the behavioral response to central oxytocin administration. J Neurosci 11:2032–2038

Winslow JT, Shapiro L, Carter CS, Insel TR 1993a Oxytocin and complex social behavior: species comparisons. Psychopharmacol Bull 29:409–414

Winslow J, Hastings N, Carter C, Harbaugh C, Insel T 1993b A role for central vasopressin in pair bonding in monogamous prairie voles. Nature 365:545–548

Winslow JT, Hearn EF, Ferguson J, Young LJ, Matzuk MM, Insel TR 2000 Infant vocalization, adult aggression, and fear behavior of an oxytocin null mutant mouse. Horm Behav 37:145–155

Young LJ, Winslow JT, Nilsen R, Insel TR 1997 Species differences in V1a Receptor gene expression in monogamous and nonmonogamous voles: behavioral consequences. 111: 599–605

DISCUSSION

Dulac: You focused on vasopressin receptor 1A. What about 1B?

Ferris: The first data I came across in terms of V1B's role in aggression was work from the Blanchards presented as a poster at Neuroscience. A number of pharmaceutical companies have worked diligently to develop ligands to block the V1A and V1B receptors for clinical use in multiple indication included aggression, anxiety and depression.

Dulac: Is this both V1A and V1B?

R *Blanchard:* Just with V1B.

Pfaff: Larry Young, at Emory University, has made a major point about the genetic manipulation of VP1A receptors as fostering affiliative behaviours. These are more-or-less the opposite of aggressive behaviours. It puzzles me: do you have explanations for his results, or perhaps alternate interpretations?

Ferris: The work you are referring to is a wonderful body of science done on monogamous and polygamous voles. Male prairie voles pair bond for life, become good mates, and help in the care of the young. They are not particularly aggressive prior to mating; however, once they pair bond the males are very aggressive towards conspecifics that come into the environment. The other phenotype is the montane vole, which is promiscuous, doesn't pair bond and is not a good parent. If one looks at the oxytocin and vasopressin receptor binding profile between these two species of voles they are strikingly different. Do these differences account for the behaviour? Winslow, Insel, Young and colleagues have manipulated the behaviour of these different voles by microinjections and transfection studies. They discovered that vasopressin could promote pair bonding, parental care and aggression in prairie voles but not montane voles. They have cloned the V1A receptor in the vole, and discovered a microsatellite repeat in the promoter region that seems to be involved in the expression of this receptor. The prairie vole has a high expression of vasopressin receptor in the ventral pallidum a part of the reward pathway. They hypothesize that with the first reproductive experience, vasopressin is being released in the ventral pallidum reinforcing the pair bonding behaviour. They put the prairie vole V1A gene into the montane vole and reported increased paternal and pair bonding behaviour (Lim et al 2004). With the increase in partner preference comes an increase in aggression toward male conpsecifics.

Martinez: I have a question related to the cortisol levels. You found a low response in adolescents that were abused and then exposed to a conflict situation. Did you measure basal levels?

Ferris: Yes basal levels of stress hormone appear to be normal between hamsters with a history of early abuse vs. sibling controls with no such history. However, when we put them in a conflict situation where they are attacked and threatened by a bigger, older male, cortisol levels do not change in early abused animals but are elevated in controls.

Brodkin: I saw a recent paper about a V1B knockout that showed reduced aggressive behaviour (Wersinger et al 2002).

Dulac: That's the paper by Larry Young, and this is why I asked the earlier question. What happens in the V1A knockout?

Ferris: Larry Young is sitting on that one. All he reported is that these animals are not anxious at all. There is a knockout of social memory.

Pfaff: Dr Dulac, is there an abundance of V1A receptors in the olfactory bulb or accessory bulb?

Dulac: I don't know. I was just curious.

Suomi: You presented some pretty strong arguments for a period of invulnerability around adolescence, with one or two exceptions. But those were animals with normal rearing histories. What do you think might happen in an individual with a compromised rearing history?

Ferris: I have thought about how we might model something like this. What if I took mom away and threw in a nasty intruder? Would the intruder attack and threaten the pups? The only problem with this possible model is the intruder doesn't touch the pups.

Dulac: What about the model with the mother that licks them less?

Suomi: That finding comes from studies with rats.

Ferris: I don't know of anyone who has looked at this in hamsters. When I first started doing immunocytochemistry in the hamster, I found out that it was an odd-looking vasopressin system. Where is all the vasopressin in the septum and amygdala? I thought I was doing something wrong. But when I spoke to others, it turned out they had been sitting on the same data for years. Vasopressin in the hamster looks nothing like the mouse or rat.

Keverne: One of the things that has puzzled me about the septum's role in aggressive behaviour is the output. Where in the brain is the output from the septum primarily influencing aggressive behaviour?

Ferris: We've looked at connections to the anterior hypothalamus through track tracing. The septum innervates the anterior hypothalamus. Activating and blocking the septum can alter agonistic behaviour. However, if you block the anterior hypothalamus you remove the effect of the septum on aggression in a hamster. Hence there is a hierarchy of brain substrates controlling aggression with the anterior hypothalamus in a primary position. I can go to the septum and get effects, but they will not supersede what is happening in the anterior hypothalamus.

Pfaff: I am quite sure that you will see that there are efferents from the septum down the medial forebrain bundle, and through the diagonal bands of Broca there will be efferents over to the amygdala. Whether or not they have enough causal power to answer your question, I don't know.

Keverne: Are there any efferents down to the ventral pallidum?

Pfaff: You would think that would be hard to avoid, but I don't know this for a fact. The person to look up is Laszlo Zaborszky.

Keverne: One of the things I noticed on your autoradiogram of the V1B receptor, is that it looked to me as if you had them in the ventral pallidum or nucleus accumbens.

C Blanchard: What about the nucleus circularis?

Ferris: We used suicide transport lectins, injected into the neurohypophysis of the pituitary gland. The lectin was carried back to all cells that projected down to the neurohypophysis involved in the release of vasopressin as a neurohormone. These cells were destroyed. When you look at the immunocytochemistry what is left is the medial supraoptic area, the nucleus circularis and a few cells in the PVN. This was our way of separating out the different populations.

References

Lim MM, Wang Z, Olazabal DE, Ren X, Terwilliger EF, Young LJ 2004 Enhanced partner preference in a promiscuous species by manipulating the expression of a single gene. Nature 429:754–757

Wersinger SR, Ginns EI, O'Carroll A-M, Lolait SJ, Young WS 2002 Vasopressin 1b receptor knockout reduces aggressive behaviour in male mice. Mol Psychiatry 7:975–984

Typology of human aggression and its biological control

Manuela Martinez and Concepción Blasco-Ros

Department of Psychobiology, Faculty of Psychology, University of Valencia, Avda Blasco Ibañez, 21, 46010 Valencia, Spain

Abstract. Human aggression is considered a global public health problem with a tremendous impact on the victims. Scientists studying this behaviour can contribute to reducing this worldwide problem by identifying its causes and designing preventive interventions and treatment. Those working on its biological bases need to establish which types of human aggression may be candidates for a biological intervention, which makes it necessary to develop a typology of this behaviour. Considering the primary goal that guides the perpetrator to behave aggressively, a distinction is made between impulsive and premeditated types. However, to control aggression, the distinction should be made between individuals instead of acts of aggression. Thus, research is being conducted to find biological markers that could identify those individuals who are at risk of behaving aggressively and in which way. The information obtained from biological studies should form a whole with that proceeding from other disciplines, such as psychology and sociology, in order to build a complete multidisciplinary picture of specific subtypes of offenders. Finally, this knowledge needs to be conveyed to policymakers, practitioners and the public in general so as to work together to develop effective strategies to address this human problem.

2005 Molecular mechanisms influencing aggressive behaviours. Wiley, Chichester (Novartis Foundation Symposium 268) p 201–215

The World Health Organization (WHO) has recently published a report on human aggression, which focused on the impact that this behaviour has on the victims, considering it as a global public health problem (Krug et al 2002). The impact of aggression is tremendous, ranging from the death of the victims to the impairment of their development and mental, physical and social health. However, while deaths represent the tip of the iceberg, most acts of aggression do not cause injuries severe enough to require medical attention and, consequently, are not reported. Apart from the high personal impact of aggression on the victims, this is not only considered a private concern, but also a societal and worldwide problem for the economy and development of countries. Human aggression costs states large quantities of money each year in healthcare, legal costs and loss of productivity, reaching in some countries 5% of the gross domestic product. Thus, the incidence of aggression in human society is so huge that it is more than enough to encourage scientists to get involved. But, how can scientists contribute to ameliorate this worldwide problem? They should work in order to understand this human phenomenon, with the final aim of identifying its different causes and, then, apply appropriate preventive interventions and treatments to reduce it. More specifically, those scientists involved in the biological bases of aggression need to determine which types of this human behaviour may be related to biological features.

Definition of human aggression

Any discussion on human aggression necessarily begins with a definition. This is a very difficult task as this behaviour is heterogeneous in its manifestations, which makes it very difficult to cover the entire spectrum in a sole definition. However, a very well accepted definition is the following: any behaviour directed toward another individual that is carried out with the immediate intent to cause harm (Geen 1990, Berkowitz 1993). In addition, the perpetrator must believe that the behaviour will harm the target and that the target is motivated to avoid the behaviour. In this definition, while the intention to harm is a necessary feature as a proximate goal, no mention is made of the primary goals. A more detailed definition has been provided by the WHO (Krug et al 2002) which defines it as: 'the intentional use of physical force or power, threatened or actual, against oneself, another person, or against a group or community, which either results in or has a high likelihood of resulting in injury, death, psychological harm, maldevelopment or deprivation'. This definition associates intentionality with the committing of the act itself, irrespective of the outcome it produces. The use of 'power' also serves to include neglect or acts of omission, in addition to the more obvious aggressive acts of commission.

Typology of human aggression

It is necessary to develop a typology of human aggression that qualitatively characterizes the different classes and the links between them. To date, research into types of human aggression has been rather limited with few general typologies existing. Most of them focus on a part of the whole phenomenon, such as the aggressive act, type of victim or specific combinations. However, a comprehensive typology needs to be based on different criteria that simultaneously cover the whole spectrum. There are many possibilities, some of which may be based on:

- the mode of the aggressive act,
- who the perpetrator is,
- the motivation that guides the perpetrator to carry out the aggressive act, and
- the characteristics of the victim.

Nevertheless, the more specific the classification of aggression the better the possibility of measuring and diagnosing each specific type more accurately. Combinations of the different criteria can be made, although not all possibilities exist.

- (a) Depending on the mode of the aggressive act, a distinction can be made between:
    - physical,
    - sexual,
    - psychological, and
    - neglect/omission.

This classification is very objective and information about each subtype can be completed by the frequency, intensity and duration. The latter subtype (neglect/omission) has been very recently incorporated, as it is aggression committed by the omission of some necessary acts.

- (b) Focusing on who the perpetrator is, aggression can be classified into the following categories:
    - self-directed, in which a person inflicts aggression upon him or herself (e.g. self-injurious behaviour, suicide)
    - interpersonal, in which the aggressive act is inflicted by another individual or by a small group of individuals (e.g. partner abuse, child abuse, rape), and
    - collective, which is usually inflicted by larger groups such as states, organized political groups or terrorist organizations (e.g. war, terrorism).

Additionally, the criteria of age and gender of the perpetrator can also be used. Regarding age, aggression can be classified into child, youth or juvenile, and adult aggression. Finally, a distinction can be made between male and female aggression.

- (c) Focusing on the primary goal and the consciousness controlling the behaviour that guides the perpetrator, a distinction, proposed by Barratt (1991), can be made between:
    - impulsive (also called emotional, reactive, hostile, affective),

- ○ premeditated (also called instrumental, proactive, predatory), and
- ○ medically related aggression, which is part of a pathological condition.

The dichotomy between impulsive and premeditated aggression has emerged as the most promising construct, although there are some controversies (Bushman & Anderson 2001). The difference between these two subtypes of aggression can be made based on three aspects:

- the primary goal,
- the presence of anger, and
- the extent of thought and planning involved.

Impulsive aggression is committed with the desire to hurt someone, is considered reactive in nature, and is accompanied by an agitated or irritated mood ('hot blooded'), with poor modulation of physiological arousal, and with a loss of behavioural control. On the other hand, premeditated aggression is motivated by some goal other than to hurt someone (e.g. obtain money), is not related to an agitated state or preceded by a potent affective reaction ('cold blooded'), is executed with low autonomic arousal, and is a planned or conscious behaviour. Finally, medically related aggression is characterized as being a symptom secondary to a medical condition including psychiatric (e.g. antisocial personality disorder, psychopathy), neurological (traumatic brain injury) or others. Some combinations of these three subtypes of aggressive categories exist. For example, impulsive aggression is commonly associated with personality disorders, while premeditated aggression has been associated with psychopathy (Woodworth & Porter 2002).

- (d) Finally, based on the characteristics of the victim, two criteria can be used: one, when the victim is related to the perpetrator, and the other, when the victim is classified by some self characteristics such as age, sex, race, etc. In the first case, a distinction can be made between:
  - ○ oneself,
  - ○ family/partner (e.g. child, partner).
  - ○ community (acquaintance, stranger).

In the second case, following the characteristic of age of the victim a distinction can be made between child, youth, adult and elder. Many other possibilities exist.

Identification of the different types of aggression

Instruments to identify the different types of aggression are needed and they should be as objective as possible. This does not pose any problem when dealing with criteria (a), (b) and (d), as the information can be assessed with very objective questions (e.g. the Conflict Tactic Scale; Straus et al 1996), but it is more difficult when dealing with the criteria 'motivation of the perpetrator'. For this purpose, different instruments for structured and semi-structured interviews or self-reports have been constructed. Most of them focus on the distinction between impulsive and premeditated aggression, such as the Impulsive/Premeditated Aggression Scales (Stanford et al 2003), the Barratt Impulsiveness Scale (Patton et al 1995), the Lifetime History of Impulsive Behaviors Interview (Schmidt et al 2004) and the Laboratory and Psychometric Measurements of Impulsivity (Cherek et al 1997). Other instruments such as the Buss-Perry Aggression Questionnaire (Buss & Perry 1992) assess aggression in a more general way.

Types of offenders and their biological markers

Scientists who work to identify the biological underpinnings of human differences in aggressive behaviour are more interested in the individual as a whole rather than in specific aggressive acts. This is necessary in order to diagnose any biological disturbance, which would increase the likelihood of an individual engaging in aggressive behaviour. Thus, scientists need to deal with types of offenders instead of types of aggressive acts. However, is it possible to move from the classification of human aggression to the classification of aggressive individuals? For example, is an individual who carries out aggressive behaviour always guided by the same type of motivation (Barratt & Felthous 2003)? If so, offenders who commit acts of impulsive aggression should be distinguished from those who commit acts of premeditated aggression. Additionally, it is necessary to determine whether this classification would be useful for all ages at which an individual may behave aggressively (e.g. childhood, youth, adulthood) (Dodge et al 1997, Vitiello & Stoff 1997), in both males and females (Connor et al 2003), in all modes of aggressive acts (e.g. physical, psychological, sexual), and towards all types of victims (e.g. partner, Chase et al 2001). Another important aspect is whether this typology can be applied both to normal, healthy individuals and to those whose aggressiveness is a symptom of a medical condition.

It would be very useful to find biological markers that could make it possible to identify each type of aggressive individual, so as to specifically intervene in order to prevent their behaviour or to control it. For example, can the different types of offenders be distinguished on the bases of molecular differences such as neurochemical brain functioning or gene polymorphism? To date, although research into types of offenders has been rather limited, efforts are continuing to determine the diagnostic signs that would make it possible to identify those of each subtype. This will eventually help with the development of strategies for early prediction, and specific interventions.

Neurochemical and genetic markers

Most research on the biological markers of aggressive individuals have focused on impulsive offenders, ranging from genetics to brain functioning. However, it would be important to study whether these markers could make it possible to distinguish between impulsive and premeditated offenders and, thus, help to predict their response to specific treatments, either from biological, psychological or social approaches.

Among the most robust findings in neuropsychopharmacology research is the relationship between brain serotonin (5-HT) function and impulsive aggression, low activity having being found to be a marker and predictor of aggression in diverse clinical, forensic and non-patient populations (Lee & Coccaro 2001). In general, impulsive offenders carrying out either self-inflicted or interpersonal aggression have low cerebrospinal fluid 5-hydroxyindolacetic acid (5-HIAA) and show a blunted hormonal response to pharmaco-challenge with agents that enhance serotonergic activity when compared to premeditated aggressors (Linnoila et al 1983, Cremniter et al 1999). Additionally, functional neuroimaging has helped to localize abnormal serotonin functioning in the brain of individuals with impulsive aggression. A reduced serotonin-mediated activation of the prefrontal cortex, indicating a dysfunction of the orbital/medial prefrontal circuit, has been found in patients with impulsive aggression in comparison to controls (New et al 2002). However, it would be important to establish whether this biological marker is a characteristic of only chronic aggressive individuals or also of those whose display of aggression is episodic, at one specific period of their lives.

On the other hand, genetic predictors may eventually help with the early prediction and prevention of aggressive behaviour. Although it would be interesting to carry out research on individuals behaviourally characterized as being either predominantly impulsive or premeditated in their aggressive behaviour, to date these studies are scarce. Research has been focused on the identification of specific molecular genetic markers of impulsiveness, which includes impulsive aggression. To date, genetic polymorphism studies have found that in genes encoding for components of the serotonergic system such as the tryptophan hydroxylase (TPH) and the monoamine oxidase A enzymes, and serotonergic receptors, specific alleles are related to impulsive aggression (New et al 1998, Manuck et al 2000). For example, TPH A218C genotypes have been associated with impulsive aggression. However, while a relation has been found between the presence of the 218C allele and impulsive aggression in well-characterized impulsive patients, this was not observed in healthy controls (Staner et al 2002).

In conclusion, identification of different types of offenders and their specific biological markers can increase the likelihood of finding specific effective psychopharmacological or psychosocial treatments for each subtype. For example, it has been postulated that psychopharmacological treatment is effective in reducing impulsive aggression, but may not reduce premeditated aggression (e.g. in incarcerated inmates, Barratt et al 1997a; in children and youths, Malone et al 1998, Steiner et al 2003). However, as individual behaviour is the result of specific interactions between biological, psychological and social variables, we should not expect to find a single marker to identify each subtype. Thus, the information obtained from biological studies should form a whole with that proceeding from other disciplines, such as psychology and sociology, in order to build a complete multidisciplinary picture of specific subtypes of offenders (Barratt et al 1997b, Dolan et al 2002, Stanford et al 2003). Finally, the progress made in research and science with respect to the classification of human offenders and their specific characteristics at biological, psychological and social levels need to be conveyed to practitioners, policymakers, and the public in order to develop effective strategies to address this human problem.

References

Barratt ES 1991 Measuring and predicting aggression within the context of personality theory. J Neuropscyhiatry Clin Neurosci 3:S35–S53

Barratt ES, Felthous AR 2003 Impulsive versus premeditated aggression: implications for Mens Rea decisions. Behav Sci Law 21:619–630

Barratt ES, Standford MS, Felthous AR Kent TA 1997a The effects of phenytoin on impulsive and premeditated aggression: a controlled study. J Clin Psychopharmacol 17:341–349

Barratt ES, Standford MS, Kent TA Felthous A 1997b Neuropsychological and cognitive psychophysiological substrates of impulsive aggression. Biol Psychiatry 41:1045–1061

Berkowitz L 1993 Aggression: its causes, consequences and control. McGraw-Hill, Inc

Bushman BJ, Anderson CA 2001 Is it time to pull the plug on the hostile versus instrumental aggression dichotomy? Psychol Rev 108:273–279

Buss AH, Perry M 1992 The aggression questionnaire. J Pers Soc Psychol 63:452–459

Chase KA, O'Leary KD, Heyman RE 2001 Categorizing partner-violent men within the reactive-proactive typology model. J Consult Clin Psychol 69:567–572

Cherek DR, Moeller FG, Dougherty DM, Rhoades H 1997 Studies of violent and non-violent male parolees: II. Laboratory and psychometric measurements of impulsivity. Biol Psychiatry 41:523–529

Connor DF, Steingard RJ, Anderson JJ, Melloni RHJr 2003 Gender differences in reactive and proactive aggression. Child Psychiatry Hum Dev 33:279–294

Cremniter D, Jamain S, Kollenbach K et al 1999 CSF 5-HIAA levels are lower in impulsive as compared to nonimpulsive violent suicide attempters and control subjects. Biol Psychiatry 45:1572–1579

Dodge KA, Lochman JE, Harnish JD, Bates JE, Pettit GS 1997 Reactive and proactive aggression in school children and psychiatrically impaired chronically assaultive youth. J Abnorm Psychol 106:37–51

Dolan M, Deakin WJF, Roberts N, Anderson I 2002 Serotonergic and cognitive impairment in impulsive aggressive personality disordered offenders: are there implications for treatment? Psychol Med 32:105–117

Geen RG 1990 Human aggression. Open University Press, Buckingham

Krug EG, Dahlberg LL, Mercy JA, Zwi AB, Lozano R 2002 World report on violence and health. World Health Organization, Geneva

Lee R, Coccaro E 2001 The neuropscyhopharmacology of criminality and aggression. Can J Psychiatry 46:35–44

Linnoila M, Virkkunen M, Scheinin M, Nuutila A, Rimon R, Goodwin FK 1983 Low cerebrospinal fluid 5-hydroxyindolacetic acid concentration differentiates impulsive from nonimpulsive violent behaviour. Life Sci 33:2609–2614

Malone RP, Bennet DS, Luebbert JF et al 1998 Aggression classification and treatment response. Psychopharmacol Bull 34:41–45

Manuck SB, Flory JD, Ferrell RE, Mann JJ and Muldoon MF 2000 A regulatory polymorphism of the monoamine oxidase-A gene may be associated with variability in aggression, impulsivity, and central nervous system serotonergic responsibility. Psychiatric Res 95:9–23

New AS, Hazlett EA, Buchsbaum MS et al 2002 Blunted prefrontal cortical ^{18}Fluorodeoxyglucose positron-emission tomography response to meta-chlorophenylpiperazine in impulsive aggression. Arch Gen Psychiatry 59:621–629

New AS, Gelernter J, Yovell Y et al 1998 Tryptophan hydroxylase genotype is associated with impulsive-aggression measures: a preliminary study. Am J Med Genet 81:13–17

Patton JH, Stanford MS, Barratt ES 1995 Factor structure of the Barratt Impulsiveness Scale. J Clin Psychol 51:768–774

Schmidt CA, Fallon AE, Coccaro EF 2004 Assessment of behavioral and cognitive impulsivity: development and validation of the Lifetime History of Impulsive Behaviors Interview. Psychiatry Res 126:107–121

Staner L, Uyanik G, Correa H et al 2002 A dimensional impulsive-aggressive phenotype is associated with the A218C polymorphism of the tryptophan hydroxylase gene: a pilot study in well-characterized impulsive inpatients. Am J Med Genet 114:553–557

Stanford MS, Houston RJ, Mathias CW, Villemarette-Pittman NR, Helfritz LE, Conklin SM 2003 Characterizing aggressive behaviour. Assessment 10:183–190

Steiner H, Saxena K, Chang K 2003 Psychopharmacologic strategies for the treatment of aggression in juveniles. CNS Spectrums 8:298–308

Straus MA, Hamby SL, Boney-McCoy S, Sugarman DB 1996 The revised Conflict Tactics Scales (CTS2). J Family Issues 17:283–316

Vitiello B, Stoff DM 1997 Subtypes of aggression and their relevance to child psychiatry. J Am Scad Child Adolesc Psychiatry 36:307–315

Woodworth M, Porter S 2002 In cold blood: characteristics of criminal homicides as a function of psychopathy. J Abnorm Psychol 111:436–445

DISCUSSION

Pfaff: Ted Brodkin, you see patients: what comments might you have?

Brodkin: I like the fact that we have been reminded again about the distinction between impulsive and premeditated aggression. A lot of this meeting has focused on impulsive aggression, and I wonder whether this is because it is easier to study in animal models, and because it is politically less charged. Premeditated aggression is also a very important phenomenon, and it is seen in clinical and non-clinical settings. In this conference on aggressive behaviour, taking place in 2004, it seems to me that the 800 lb gorilla sitting in the room that we haven't mentioned yet is the issue of war and terrorism. One can look at war or terrorism as carefully planned, premeditated types of aggression. I am not saying there is a biological solution for these huge issues, but they are issues worth bringing up here.

Gammie: Scott Atran from the University of Michigan is considered one of the leading experts on terrorism. He published a review article a few years ago (Atran 2003) in which he argued that maternal aggression was one of the best models. He felt that suicide bombers were performing something equivalent to maternal aggression. When he interviewed these people most of them weren't thugs, and they felt like their kin were under the gun. They felt this was an act of bravery where they were protecting what they considered their family group. If you look at the Palestinian situation, it seems like they are under the gun and they are exhibiting something akin to maternal defence-type behaviour.

Hinde: I thought this was an interesting classification but ultimately we are going to need a multidimensional one. I have been trying to relate your classification into impulsive and premeditating to the offensive–defensive dichotomy that we have been dealing with so far. It seems that if you take the old functional classifications that have been used for human aggression, things like instrumental aggression can be either, and spontaneous aggression when it is like teasing can be either. We need to look at both of these dimensions.

I would also like to say a word about prevention of aggression. You talked mainly about treatment of people who already showed aggressive behaviour. I believe that perhaps the most important issue in giving rise to aggression in humans is the mother–child relationship. There is a lot of evidence for this now. What the data seem to show is that authoritarian parents and very neglectful parents give rise to aggressive children. Sensitive parents who exercise firm control give rise to less aggressive children. There are of course many social and environmental issues that come in after this. If we want to reduce aggression we have to concentrate on two things. One is poverty and the culture of aggression that is often a consequence of poverty. Aggressive acts have a whole string of causes behind them, and poverty is one of those. The culture and social environment is also critical in determining whether aggression is acceptable. The second thing we need is education, and especially education of women. I am an idealist and I want to put the world to rights! These two things would go a long way to ameliorating the problem of violence.

Pfaff: In terms of preventive strategies you might be interested by a book by James Gilligan, a prison psychiatrist who is a professor at Harvard, entitled *Preventing Violence (Prospects for Tomorrow)* (Gilligan 2001). Also, the New York Academy of Science held a conference this year on strategies for the prevention of youth violence (*http://www.nyas.org/ebriefreps/main.asp?intSubsectionID=695*). A tremendous percentage of violent acts are committed by young males. Finally, the education of women is also key in stabilizing or reducing world population.

Hinde: Overpopulation is one of the causes of aggression. But the strongest correlation of violence both within societies and between societies is the disparity of wealth between rich and poor.

Skuse: I wanted to take Manuela back to a discussion we had over lunch, which concerned the distinction between violence and aggression. If we define violence the way that the WHO does, this is clearly largely perpetrated by males. We are looking at a male problem. It is interesting that the vast majority of people in this room are males, yet we haven't really focused on why it is males who perpetrate so much violence.

Suomi: There is an entire line of research focused on what Nicki Crick calls 'relational aggression', shown primarily by females. This relationship aggression shows the same kinds of developmental changes seen in male physical aggression and is subject to many of the same variables. It has the same relationship with a variety of other variables.

Skuse: This is touching on what we were talking about: female aggression manifests in a different way but is not necessarily any less common than male aggression (Potegal & Archer 2004).

Martinez: Some researchers consider that male and female aggression are similar in quantity. Professor Murray A. Straus, from the University of New Hampshire, says that in intimate partner violence, women are as aggressive as men. I don't totally agree with this.

Skuse: It depends on how you want to define it. If you define aggression in terms of physical acts of aggression, then it is perpetrated mainly by males. If you are talking about a more subtle form of aggression in the way that one tends to manipulate someone else, it may be equal.

Martinez: In the case I was referring to, Professor Straus was studying physical aggression. However, there are specific types of aggression, such as indirect aggression, that are considered a typical female behaviour. But, considering human aggression as a problem, I don't think we can put aggression perpetrated by men and that perpetrated by women on the same level.

Ferris: If you go back to your typology you'll see that young males account for 95% of all the violence.

Pfaff: Martin Daly and Margot Wilson, evolutionary psychologists in Canada, have a lot of data across cultures. If one plots the murder of males by unrelated males on the time base of an entire life cycle, and then plot testosterone levels the same way, they are parallel.

C Blanchard: First of all, they are pointing out that a huge proportion of crimes of violence tend to be committed by males between 15 and 30, especially unmarried males. If this is put in an evolutionary perspective, one could make the case that those young unmarried men are aggressive in the context of acquiring things that attract women: status in a peer group or gang, or money. Daly and Wilson regard this as a mate or resource seeking strategy. In our discussions I haven't heard the idea that aggression has functions. In an evolutionary perspective, it does have functions. These are pretty clear in infrahuman mammals. In people the picture is fuzzier because there are so many cultural and legal sanctions and considerations. However, even if you look at individual fighting, a generally nonapproved form of human aggression, it tends to involve people who have something to gain from it, or feel they have something to lose if they don't do it.

Skuse: Should we equate aggression and violence?

Hinde: Absolutely not. War is violence, but the violence in war is seldom caused by aggressiveness. Bomber crews set out because it is their duty to do so; tank commanders advance because it is their duty to do so. Duty is the primary cause of violence in war. Aggressiveness may come into it, but it is very secondary.

Pfaff: What do you think about terrorism: do you think that is seen as a duty by the perpetrator?

Hinde: Yes.

R *Blanchard:* Don't you think anger is a component of terrorism? Anger provides the substrate for the acts.

Hinde: If what you are saying is that it has a whole network of interconnected causes, I agree.

Nelson: With regard to the distinction between impulsive and premeditated aggression, when we think about a lot of aggression the driving force is revenge. The decision to do something might have been impulsive, but the action premeditated.

Martinez: I have a question for the audience. We have been working on the biological mechanism of aggression. How can we contribute to solve the human problem of violence? In what proportion of cases can we be of help?

Nelson: That's a difficult question. Let's say we have a biological marker of aggression. What will you do with that information?

Hinde: I want to get back to the importance of distinguishing between aggression and aggressiveness. If you are talking about a biological marker, this is a marker for the proneness to be aggressive. The actual act of aggression has a network of causes.

Nelson: I can give you an easy example. Our group did a study where we measured impulsiveness by how hard someone hung up a telephone when we angered them in a psychological study. We brought the people in, had them spit and measured their testosterone along with other biological data. We measured length of digits, ears, eye placement, and other morphological traits with the idea that more asymmetrical individuals would be more impulsive. There is a literature showing that asymmetrical individuals had perturbations during development.

The first time the participants were in the lab, we told them that they had high testosterone, which we informed them is a sign of persuasiveness — we were misleading them, but debriefed them at the end of the study. Then we asked them to call some people: we said we would give them money if they persuaded people to give money to a charity. Of course, they were going to fail because we had actors on the phone. We measured how hard they hung the phone up. They then had to send the people they had spoken to letters. We found that people who were asymmetrical in their traits were much more impulsive and likely to send a nasty letter (Benderlioglu et al 2004). The reviews we got on this paper were dramatic. There was a very strong negative response because we were trying to find these markers. We were accused of engaging in 'social engineering'. People aren't ready for biological markers of these kinds of traits.

Martinez: People are afraid of finding biological markers because they think once that you have found one, the only way to intervene to prevent aggression is by providing biological treatment. However, there are other options than giving people tablets or manipulating them genetically, for example, social interventions.

Nelson: I agree, but the fact that the 'peers' who reviewed this paper were put off indicates that the general public will probably be less receptive to these ideas.

Craig: You had suicide built into this. I have a problem, which is with the cross-over between depression and aggression. It seems that suicide is the end point of extreme depression, yet you are building this into a violent kind of category.

Martinez: There is a distinction between impulsive and premeditated suicide. And, although it has been found that depression can lead to suicide, not all cases are related to depression. Furthermore, suicide is not a behaviour that occurs in a vacuum but in a social context.

Craig: Is that aggression? It doesn't seem to me to fit into any category.

Koolhaas: There is a high comorbidity between depression and aggression.

Manuck: I can't help thinking that we are a little out of our depth discussing what we are going to do about the 'problem' of aggression. I would second Dr Blanchard's point: we should think perhaps first about the functions that aggression serves, and then maybe about the varieties of aggressive action that in their extreme bring some people to forensic or clinical attention. With respect to clinically recognized aggression, men are more frequently diagnosed with antisocial personality disorder (ASPD) than women. Prominent ASPD symptoms include violations of law and social norms, irritability and aggressiveness (physical assaults), lack of remorse over the mistreatment of others, and patterns of irresponsibility, recklessness, and impulsivity. In addition, we now recognize that the ASPD 'label' may subsume aggressive and antisocial behaviours of differing aspect and aetiology. In addition to people whose aggression stems from poorly regulated impulse and affect, others simply use aggression to get what they want (instrumental aggression), unimpeded by conscience or socialization due to an impaired capacity for emotional learning (Blair 2003). The latter category comprises perhaps 30% of diagnosed ASPD, and the behaviour of these individuals is likely mediated by psychological and neural mechanisms different from those of impulsive aggression (Hart & Hare 1997, Blair 2001). How these clinical phenotypes relate to normal behavioural variation is unclear. It seems likely that impulsive aggression is coextensive with normative variability in antagonistic disposition and impulse control; in contrast, perhaps true psychopathy—the cold (unfeeling) aggression of instrumental motivation—is discontinuous with traits commonly distributed in the population, and therefore less easily understood as the end of a continuum of normal variation.

Suomi: I want to mention the work of Richard Tremblay in Montreal. Over the last 15 years he has carried out some groundbreaking work in the general area of the development of aggression, and he has devised some heroic intervention approaches. He has carried out longitudinal studies of youths and has shown that the highest incidence of aggression as a proportion of total behaviour is among 2-year-olds. In most cases aggression starts declining after that, but for some individuals it does not. He has described four different developmental trajectories, the most relevant of which is the group that starts high and remains high throughout the rest of development. The stability of the individual differences on these measures of aggression throughout the whole of development is remarkable. On the basis of these findings he initiated an intervention program targeting 4-, 5-, and 6-year-old children. It involves educating parents and teachers with respect to issues of insults and aggressiveness, and certain activities designed to promote peer pressure. These are universal interventions that are basically done at the classroom level. Tremblay has impressive results over a two- or three-year period. He is now about to take this intervention program into all primary schools in Quebec.

Ferris: When can you start working with these children? You have to wait until they get to school. Tremblay started these interventions at fourth grade and they didn't work, so he went down to second and first grade. The earlier you get to these children the more effective the program is.

Suomi: The nice thing about Tremblay's program is it as much preventive as it is an intervention program, and it is a universal program so it is not targeting particular individuals.

R Blanchard: It is amazing to me that the problems we talked of before, the problem of social control and the threat of stigmatization, are so central. When we look at the animal literature we are seeing a clear type of aggression which is a real phenomenon, such as male offensive or female offensive behaviours. These types of behaviour are clearly much more closely related to impulsive aggression in the human dimension. I am not sure I can see an animal model of premeditation.

The other type we see in animals is a variety of aggression that is seldom mentioned in the human literature, which is fear-based aggression. This is defensive. We know that a number of murders in the USA are committed by people who are essentially afraid and who have weapons in their hands. To what extent do you feel that there is a fear-based defensive aggression in humans?

Martinez: I agree that there is a fear-based defensive aggression in humans. However, the perception of dangerous signals that induce fear may vary depending on the individual and the context. For example, some studies in children have shown that those who show reactive impulsive aggression have problems in perceiving the signals from their peers. They identify the normal behaviour of their schoolmates as dangerous for them so they react aggressively.

R Blanchard: Defensive aggression is very noisy, at least in animals, whereas offensive aggression is quiet. These are quite distinct in the animal literature, and I have yet to see the dichotomy being fully examined in humans. Bullying is probably related to what we call impulsive, dominant-type angry aggression.

Martinez: In my opinion, the picture of the typology of human aggression is more complete day by day. But at the moment there has not been established a relationship between all types of human aggression and the function of the specific behaviours.

Ferris: Robert Blanchard, you asked earlier whether we could medicalize the problem of aggression. If we could, then we could focus on the biology. However, the impulsivity and violence falling under this rubric would be such a small subset of the much greater problem of societal violence. Manuela Martinez, in your typology one of the things I liked was your attention to the victim. One constant in all the permutations of violence is the victim. How much of the health dollar is going to the victim in terms of their treatment?

Martinez: This is an important aspect in the field of aggression in my opinion. The cost of the treatment of the victim is very high. However, this is not only a burden for the health system but also for society in general, as it involves social services, the judicial system, care of children, etc, which can be very high in rich countries. For example, the cost of intimate partner violence has been calculated in some countries to be as high as 5% of the Gross Domestic Product.

Hinde: An extensive study of bullying in schools by Olweus in Norway showed that there were specific characteristics of the boys who were bullied (Olweus 1978). He calls these the victim characteristics, and these were almost as clear cut as the characteristics of the bullies.

References

Atran S 2003 Genesis of suicide terrorism. Science 299:1534–1539

Benderlioglu Z, Sciulli P, Nelson RJ 2004 Fluctuating asymmetry predicts human reactive aggression. Am J Human Biol 16:458–469

Blair RJR 2001 Neurocognitive models of aggression, the antisocial personality disorders, and psychopathy. J Neurol Neurosurg Psychiatry 41:727–731
Blair RJR 2003 Neurobiological basis of psychopathology. Brit J Psychiatry 182:5–7
Gilligan J 2001 Preventing violence: prospects for tomorrow, Thames & Hudson
Hart SD, Hare RD 1997 Psychopathy: Assessment and association with criminal conduct. In: DM Stoff, J Breiling, JD Maser (Eds) Handbook of antisocial behavior. Wiley, New York, p 22–35
Olweus D 1978 Aggression in the schools: bullies and whipping boys. Hemisphere (Wiley), Washington DC
Potegal M, Archer J 2004 Sex differences in childhood anger and aggression. Child Adolesc Psychiatr Clin N Am 13:513–28, vi–vii

Aggression and social behaviour in rhesus monkeys

Stephen J. Suomi

Laboratory of Comparative Ethology, National Institute of Child Health & Human Development, NIH, Bethesda, MD 20892-7971, USA

Abstract. Recent research has disclosed marked individual differences in biobehavioural responses to social conflicts exhibited by rhesus monkeys across the life span. For example, approximately 5–10% of rhesus monkeys growing up in the wild consistently exhibit impulsive and/or inappropriately aggressive responses to mildly stressful situations throughout development; those same individuals also show chronic deficits in their central serotonin metabolism. These characteristic patterns of biobehavioural response emerge early in life and remain remarkably stable from infancy to adulthood. Laboratory studies have demonstrated that although these characteristics are highly heritable, they are also subject to major modification by specific early experiences, particularly those involving early social attachment relationships. Moreover, genetic and early experience factors can interact, often in dramatic fashion. For example, a specific polymorphism in the serotonin transporter gene is associated with deficits in early neurobehavioural functioning and serotonin metabolism, extreme aggression, and excessive alcohol consumption among monkeys who experienced insecure early attachment relationships but not in monkeys who developed secure attachment relationships with their mothers during infancy. Because daughters tend to develop the same type of attachment relationships with their own offspring that they experienced with their mothers early in life, such early experiences provide a possible non-genetic mechanism for transmitting these patterns to subsequent generations.

2005 Molecular mechanisms influencing aggressive behaviours. Wiley, Chichester (Novartis Foundation Symposium 268) p 216–226

This chapter describes a program of research investigating the behavioural and biological basis of individual differences in the expression of aggression among rhesus monkeys growing up in both naturalistic and captive settings. It focuses on a subgroup of rhesus monkeys who spontaneously exhibit excessive and socially inappropriate aggression and other patterns of impulsive behaviour, as well as chronic deficits in serotonin metabolism, throughout ontogeny. The research demonstrates that both genetic and environmental factors clearly can influence the development of these behavioural and biological propensities.

Perhaps more importantly, those genetic and environmental factors may actually be interacting in shaping such development.

Ethologists have long argued that the expression of aggressive behaviour can serve important adaptive functions for social mammals, since such behaviours have largely been conserved over their evolutionary history (Lorenz 1966). The capacity to engage in aggressive attack and defence in the service of promoting and protecting self, family and friends from predators and competitors seems crucial for the survival of the individual and the maintenance of any social group across successive generations. However, excessive and/or inappropriate aggression initiated by any group member has the potential to destroy the very social fabric that binds that group together. It follows that the expression of aggression must somehow be regulated, i.e. individuals must come to know which social stimuli merit aggressive responses and which do not — and for those that do, to what degree, and for how long (Suomi 2000).

Developing the capacity to regulate aggression appears to be especially important for those advanced primate species whose members live in large social groups that are well defined in terms of both kinship relationships and social dominance hierarchies. Among the most complex are those of rhesus monkeys (*Macaca mulatta*), a highly successful species of macaque monkey that lives throughout most of the Indian subcontinent and beyond. In their natural habitats rhesus monkeys typically reside in large, distinctive social groups ('troops') composed of several female-headed families, each spanning three or more generations of kin, plus numerous immigrant adult males. This form of social group organization derives from the fact that all Rhesus monkey females spend their entire life in the troop in which they were born, whereas virtually all males emigrate from their natal troop around the time of puberty and eventually join other troops. Rhesus monkey troops are also characterized by multiple social dominance relationships, including distinctive hierarchies both between and within families, as well as a hierarchy among the immigrant adult males. These complex familial and dominance relationships seemingly require that any well-functioning troop member not only be able to regulate their expressions of aggression but also to become familiar with the specific kinship and dominance status of other monkeys toward whom they might be expressed. An impressive body of both laboratory and field data strongly suggest that the acquisition of such knowledge represents an emergent property of the species-normative pattern of socialization that Rhesus monkey infants typically experience throughout development (Sameroff & Suomi 1996).

Both laboratory and field studies have documented dramatic individual differences among Rhesus monkeys in their biobehavioural responses to social conflicts and other environmental challenges. Some monkeys consistently appear to be unusually impulsive and aggressive, often initiating severe and inappropriate attacks on others in their social group or unnecessarily provoking aggressive responses from them. Most of these impulsively aggressive individuals, comprising perhaps 5–10% of both captive and field populations of Rhesus monkeys studied to date, also exhibit apparent deficits in serotonergic function, as reflected by unusually low chronic cerebrospinal fluid (CSF) concentrations of the central serotonin metabolite 5-hydroxy-indoleacetic acid (5-HIAA) (Higley & Suomi 1996).

Many of the behavioural patterns and tendencies that distinguish impulsively aggressive monkeys from others in their social group initially appear in late infancy and remain remarkably consistent throughout the whole of development. Young males growing up in field settings who exhibit such patterns initially do so in the context of rough-and-tumble play with peers—their play tends to be excessively aggressive, often escalating inappropriately to exchanges involving actual physical attack. They also are more likely to engage in other forms of risky behaviour, e.g. taking dangerous leaps from treetop to treetop, than are their peers (Mehlman et al 1994). At later ages these same males are disproportionately likely to repeatedly confront high-ranking adult males in their social group, often with physically damaging consequences for themselves (Higley et al 1992). Individual differences in CSF 5-HIAA concentrations in Rhesus monkeys have been found to be at least as stable throughout development as are the above-described behavioural tendencies. Indeed, CSF 5-HIAA concentrations measured at 14 days of age are significantly correlated with those obtained from the same subjects at 4 years of age, as well as at intermediate ages (Higley & Suomi 1996).

Laboratory studies have demonstrated that individuals with the lowest CSF 5-HIAA concentrations are disproportionately likely to have poor state control and visual orienting capabilities during early infancy (Champoux et al 1994), and perform poorly on delay-of-gratification tasks during childhood (Bennett et al 1999). In addition, they exhibit altered sleep patterns (Zajicek et al 1997), as well as unusually high cerebral glucose metabolism (as assessed by positron emission tomography, PET) during adulthood (Doudet at al 1995). These monkeys also tend to consume excessive amounts of alcohol when placed in a 'happy hour' setting as young adults (Higley et al 1991).

In free-ranging settings Rhesus monkey males with the lowest CSF 5-HIAA are far more likely to be expelled from their natal troop prior to puberty (Mehlman et al 1995) and are less likely to survive to adulthood than are other males in their birth cohort (Higley et al 1996a). Young females who have chronically low CSF levels of 5-HIAA also tend to be impulsive, aggressive, and generally rather incompetent socially. However, unlike those males, they are not expelled from their natal troop but instead remain with their families throughout their lifetime (Westergaard et al 2003), although studies of captive Rhesus monkey groups suggest that these females usually remain at the bottom of their respective dominance hierarchies (Higley et al 1996b), and their maternal behaviour is often inadequate or even abusive (Suomi 2000). In sum, wild Rhesus monkeys who exhibit poor regulation of impulsive and aggressive behaviour and deficits in central serotonin metabolism early in life tend to follow developmental trajectories that often result in premature death for males and chronically low social dominance and poor parenting for females.

Studies of captive Rhesus monkey colonies have demonstrated that individual differences in CSF 5-HIAA concentrations are highly heritable (Higley et al 1993). On the other hand, CSF 5-HIAA concentrations, along with a host of other biobehavioural measures, are also clearly subject to major modification by early social experiences, especially those involving attachment relationships. For example, Rhesus monkeys raised from birth away from their biological mothers and other adults but in the continuous company of similarly reared age mates for their first 6 months typically develop biobehavioural profiles that seem to mimic those features characteristic of excessively aggressive monkeys observed in both laboratory settings and naturalistic habitats. Peer-reared monkeys consistently exhibit lower CSF 5-HIAA concentrations than their mother-reared counterparts throughout development, in addition to higher rates of impulsive aggression and excessive alcohol consumption in adolescence and early adulthood (Higley et al 1996c). In short, both genetic and early experiential factors can affect a monkey's characteristic pattern of biobehavioural reactivity.

Do these factors operate independently, or do they interact in some fashion in shaping individual developmental trajectories? Recent research has demonstrated several significant gene–environment (G×E) interactions between allelic variation in the serotonin transporter (5-HTT) gene and early social experiences in shaping developmental trajectories for Rhesus monkeys. This gene has length variation in its promoter region that results in allelic variation in 5-HTT expression in both humans and Rhesus monkeys (Lesch et al 1997). A heterozygous 'short' allele (LS) confers low transcriptional efficiency to the 5-HTT promoter relative to the homozygous 'long' allele (LL), raising the possibility that low 5-HTT expression may result in decreased serotonergic function (Heils et al 1996).

Several studies have now shown that the consequences of having the LS allele differ dramatically for peer-reared monkeys and their mother-reared counterparts. For example, Bennett et al (2002) found that CSF 5-HIAA concentrations did not differ as a function of 5-HTT status for mother-reared subjects, whereas among peer-reared monkeys individuals with the LS allele had significantly lower CSF 5-HIAA concentrations than those with the LL allele. Analysis of HPA responsivity to short-term social separation in mother- and peer-reared juveniles revealed a parallel G×E interaction pattern: peer-reared LS subjects showed excessive elevations in plasma adrenocorticotropic hormone (ACTH) concentrations relative to peer-reared LL and both LL and LS mother-reared subjects (Barr et al 2004a). A similar pattern appeared with respect to aggression: high levels of aggression were shown by peer-reared monkeys with the LS allele, whereas mother-reared LS monkeys exhibited low levels that were comparable to those of both mother-reared and peer-reared LL monkeys (Barr et al 2003). Champoux et al (2002) examined the relationship between early rearing history and serotonin transporter gene polymorphic status on measures of neonatal neurobehavioural development during the first month of life and found that infants possessing the LS allele who were being reared in the laboratory neonatal nursery showed significant deficits in measures of attention, activity and motor maturity relative to nursery-reared infants possessing the LL allele; in contrast both LS and LL infants who were being reared by competent mothers exhibited normal values for each of these measures.

An even more dramatic G×E interaction was revealed by an analysis of alcohol consumption data: whereas peer-reared monkeys with the LS allele consumed more alcohol than peer-reared monkeys with the LL allele, the reverse was true for mother-reared subjects, with individuals possessing the LS allele actually consuming less alcohol than their LL counterparts (Barr et al 2004b). In other words, the LS allele appeared to represent a risk factor for excessive alcohol consumption among peer-reared monkeys but a protective factor for mother-reared subjects.

It could be argued on the basis of these findings that having the LS allele of the 5-HTT gene may well lead to psychopathology among monkeys with poor early rearing histories but might actually be adaptive for monkeys who develop secure early attachment relationship with their mothers. The implications of these recent findings may be considerable with respect to the cross-generational transmission of these biobehavioural characteristics, in that the attachment style of a monkey mother is typically copied by her daughters when they grow up and become mothers themselves (Suomi 1999). If similar maternal 'buffering' were indeed experienced by the next generation of infants carrying the LS 5-HTT polymorphism, then having had their mothers develop a secure attachment relationship with their own mothers might well provide the basis for a non-genetic means of transmitting its apparently adaptive consequences to that new generation. On the other hand, if contextual factors such as changes in dominance rank, instability within the troop, or changes in the availability of food were to alter a young mother's care of her infants in ways that compromised such buffering, then one might expect any offspring carrying the LS polymorphism to develop some if not all of the problems described above.

In summary, this chapter described research focusing on a subgroup of Rhesus monkeys who spontaneously exhibit excessive and socially inappropriate aggression and other patterns of impulsive behaviour, as well as chronic deficits in serotonin metabolism, throughout ontogeny. Both genetic and environmental factors clearly can influence the development of these behavioural and biological propensities, and indeed they may actually interact in shaping such development. A case in point involves interactions between different polymorphisms of the 5-HTT gene and early social rearing experiences. Rhesus monkeys possessing the LS allele exhibited significant deficits in early neurobiological functioning and serotonin metabolism, extreme aggressiveness and HPA responsiveness, and excessive alcohol consumption relative to monkeys possessing the LL allele, but only if they had been peer-reared; no such behavioural or biological deficits were shown by LS monkeys who had been reared by competent mothers. Furthermore, because rhesus monkey mothers tend to develop the same type of early attachment relationships with their offspring that they experienced with their own mothers early in life, such experiences may provide a non-genetic mechanism for transmitting these patterns to subsequent generations of monkeys.

Acknowledgements

The research summarized in this report was supported by funds from the Division of Intramural Research, National Institute of Child Health & Human Development, National Institutes of Health, DHHS. Additional funding was provided by the DIR, NIAAA, NIH, DHHS.

References

Barr CS, Newman TK, Becker ML et al 2003 The utility of the non-human primate model for studying gene by environment interactions in behavioral research. Genes Brain Behav 2:336–340

Barr CS, Newman TK, Lindell SG et al 2004a Interaction between serotonin transporter gene variation and rearing condition in alcohol preference and consumption in female primates. Arch Gen Psychiatry 61:1146–1152

Barr CS, Newman TJ, Shannon C et al 2004b Rearing condition and 5-HTTLPR interact to influence LHPA-axis response to stress in infant macaques. Biol Psychiatry 55:733–738

Bennett AJ, Tsai T, Hopkins WD et al 1999 Early social rearing environment influences acquisition of a computerized joystick task in rhesus monkeys (*Macaca mulatta*). Am J Primatol 49:33–34

Bennett AJ, Lesch KP, Heils A et al 2002 Early experience and serotonin transporter gene variation interact to influence primate CNS function. Mol Psychiatry 17:118–122

Champoux M, Suomi SJ, Schneider ML 1994 Temperamental differences between captive Indian and Chinese-Indian hybrid rhesus macaque infants. Lab Anim Sci 44:351–357

Champoux M, Bennett AJ, Lesch KP et al 2002 Serotonin transporter gene polymorphism and neurobehavioral development in rhesus monkey neonates. Mol Psychiatry 7:1058–1063

Doudet D, Hommer D, Higley JD et al 1995 Cerebral glucose metabolism, CSF 5-HIAA, and aggressive behavior in rhesus monkeys. Am J Psychiat 152:1782–1787

Heils A, Teufel A, Petri S et al 1996 Allelic variation of human serotonin transporter gene expression. J Neurochem 6:2621–2624

Higley JD, Suomi SJ 1996 Reactivity and social competence affect individual differences in reaction to severe stress in children: Investigations using nonhuman primates. In: Pfeffer CR

(ed) Intense stress and mental disturbance in children. American Psychiatric Association Press, Washington, p 3–58
Higley JD, Hasert ML, Suomi SJ, Linnoila M 1991 A new nonhuman primate model of alcohol abuse: Effects of early experience, personality, and stress on alcohol consumption. Proc Natl Acad Sci USA 88:7261–7265
Higley JD, Mehlman PT, Taub DM et al 1992 Cerebrospinal fluid monoamine metabolite and adrenal correlates of aggression in free-ranging rhesus monkeys. Arch Gen Psychiatry 49:436–444
Higley JD, Thompson WT, Champoux M et al 1993 Paternal and maternal genetic and environmental contributions to CSF monoamine metabolites in rhesus monkeys (Macaca mulatta). Arch Gen Psychiatry 50:615–623
Higley JD, Mehlman PT, Taub DM et al 1996a Excessive mortality in young free-ranging male nonhuman primates with low CSF 5-HIAA concentrations. Arch Gen Psychiatry 53:537–543
Higley JD, King ST, Hasert MF, Champoux M, Suomi SJ, Linnoila M 1996b Stability of individual differences in serotonin function and its relationship to severe aggression and competent social behavior in rhesus macaque females. Neuropsychopharmacology 14:67–76
Higley JD, Suomi SJ, Linnoila M 1996c A nonhuman primate model of Type II alcoholism?: Part 2: Diminished social competence and excessive aggression correlates with low CSF 5-HIAA concentrations. Alcohol: Clin Exp Res 20:643–650
Lesch LP, Meyer J, Glatz K et al 1997 The 5-HT gene-linked polymorphic region (5-HTTLPR) in evolutionary perspective: alternative biallelic variation in rhesus monkeys. J Neur Trans 104:1259–1266
Lorenz K 1966 On aggression. Harcourt Brace World, New York
Mehlman PT, Higley JD, Faucher I et al 1994 Low cerebrospinal fluid 5 hydroxyindoleacetic acid concentrations are correlated with severe aggression and reduced impulse control in free-ranging primates. Am J Psychiatry 151:1485–1491
Mehlman PT, Higley JD, Faucher I et al 1995 CSF 5-HIAA concentrations are correlated with sociality and the timing of emigration in free-ranging primates. Am J Psychiatry 152:901–913
Sameroff AJ, Suomi SJ 1996 Primates and persons: a comparative developmental understanding of social organization. In: Cairns RB, Elder GH, Costello EJ (eds) Developmental science. Cambridge University Press, Cambridge, p 97–120
Suomi SJ 1999 Attachment in rhesus monkeys. In: Cassidy J, Shaver PR (eds) Handbook of attachment: theory, research, and clinical applications. Guilford Press, New York, p 181–197
Suomi SJ 2000 A biobehavioral perspective on developmental psychopathology: excessive aggression and serotonergic dysfunction in monkeys. In Sameroff AJ, Lewis M, Miller S (eds) Handbook of developmental psychopathology. Plenum Press, New York, p 237–256
Westergaard GC, Suomi SJ, Chavanne TJ et al 2003 Physiological correlates of aggression and impulsivity in free-ranging female primates. Neurophysiology 28:1045–1055
Zajicek K, Higley JD, Suomi SJ, Linnoila M 1997 Rhesus macaques with high CSF 5-HIAA concentrations exhibit early sleep onset. Psychiatry Res 77:15–25

DISCUSSION

Manuck: Your group reported recently that among animals with low CSF 5-HIAA concentrations, males that sired offspring tended to be younger than those that didn't. And conversely, among animals with high 5-HIAA levels, it was the older males that successfully sired offspring.

Suomi: That was our initial field report, and we are not finding that age difference now. Indeed, we are having trouble finding low CSF 5-HIAA males in field sites at Morgan Island.

Manuck: I believe that the age at which males emigrate from their natal groups covaries inversely with animals' CSF 5-HIAA concentrations. This was shown both at Morgan Island (Mehlman et al 1995) and Kayo Santiago (Kaplan et al 1995).

Suomi: Yes. In fact, we have an interesting situation where a different endophenotype of unusually fearful males tend to postpone emigration until their sixth, seventh or eighth year, which actually puts them at an advantage. Our best predictor of survival among emigrating males is how physically large and heavy they are at the time they leave home. Males who can postpone emigration until they finish their adolescent growth spurt are better off than males who go in the middle of their growth spurt or before it ever starts. These fearful, nervous males have a better chance of surviving than do others because they tend to emigrate later.

Manuck: So you don't necessarily agree with Berard and colleagues' report that homozygosity for the short allele of the rhesus serotonin transporter polymorphism predicts earliest dispersal and homozygosity for the long allele, latest emigration (Trefilov et al 2000).

Suomi: No, that pattern holds. I am just putting the behavioural phenotype on top of this.

Manuck: As I understand it, he was arguing that an intermediate age of emigration was optimal with respect to reproductive success and that because intermediate-age dispersal characterized the heterozygous male, this pattern might be maintained in the population through balancing selection.

Suomi: We think otherwise. Particularly at risk are the monkeys getting kicked out prior to puberty. Most of them don't survive the emigration process.

Dulac: What is the chance that a daughter being raised by a mother with a short allele and experiencing poor rearing, will in turn become a good mother?

Suomi: The nature of her experience has to do with two things. When females grow up in benevolent environments they do pretty well independent of their genotypic status. As far as what sort of mother they will turn out to be, it will be most likely what they experienced with their own mothers when they were growing up.

Dulac: So is there any way out of the vicious circle of having the wrong allele and being badly reared? In other words, why is the bad allele still in the population?

Suomi: It's time to tell some stories. There is a subgroup of rhesus monkeys who migrated from India over the Himalayas and into China. Some of these monkeys started showing up in US labs in recent years. The word on the street among lab technicians was that these Chinese-derived rhesus monkeys were really bad, nasty animals. They were even more aggressive than rhesus monkeys normally are, and they were intractable and especially difficult to train. We had the opportunity to rear some of these monkeys in our nursery, because some of my colleagues who were conducting contraceptive studies had their mothers as subjects. The contraceptives didn't always work, so they ended up with a lot of unwanted infants. They didn't have a nursery facility themselves, so they gave us the infants to rear in our nursery. Within a week my nursery staff were asking what was wrong with these infants because they were biting the staff members who were trying to feed them, something that had never happened before. These infants were irritable and difficult to handle. When we ran them through our standard neonatal test battery, we found that they performed very poorly on measures of state control and visual orienting. Their CSF 5-HIAA values were very low. We then watched them grow up, and they were as aggressive as can be. Finally, we genotyped them and found that whereas the gene frequency of the short allele in our overall rhesus monkey colony derived from India many generations ago was 23%, the incidence of the short allele in this group of Chinese-derived rhesus monkeys was 67%, i.e. the short allele was the rule rather than the exception. I can think of many reasons why this might have happened. Perhaps these were offspring of monkeys who were stupid enough to get caught in the first place, perhaps it was because they were only two generations removed from the wild, but my favourite explanation is the following: any monkey impulsive or unpopular enough to be driven over the Himalayas must be a bit special! The habitat on the Chinese side of the Himalayas is basically inhospitable and largely uninhabitable for humans. Perhaps in exceedingly adverse environments, this trait might be advantageous. Rhesus monkeys can live just about everywhere: they are a weed species. Perhaps this capability has to do with having a range of genetic material to work with, and in some circumstances this so-called 'bad' gene might actually be adaptive.

Robins: Can you detect social changes in monkeys that have made it over the mountains?

Suomi: We are trying to get someone to do a field study there.

Martinez: What interested me most about your work is that a biological predisposition can be reversed by social intervention. This is very important. When we try to find biological markers for aggressive behaviours in humans, people start trembling, because they think that the only way we can reverse this predisposition is by drugs. Another aspect in which I am interested is whether this finding could be applied to humans? For example, some children are biologically predisposed to be more aggressive than others. Would you suggest determining the levels of serotonin in these children? And in those cases in which there is a biological predisposition, would the state provide support to the families in case the mothers were not competent enough to reverse the biological predisposition?

Suomi: As I have said so many times, jumping to precise comparisons and conclusions about what we might do in humans on the basis of what happens in animals is fraught with potential problems. Rhesus monkeys are not furry little humans with tails; they are members of another species, albeit a closely related one. We have to be careful with direct one-to-one comparisons. It is incumbent on us to look for generality of basic principles across different species, however. The principal that there may be regulation of gene action by environmental events is one that is no doubt applicable across the animal kingdom. With regard to your question, the best we can probably do is support the mothers. Females who become mothers from other at-risk groups are at high risk of neglecting or abusing their offspring without social support or in an unstable social situation. In a stable social group they can turn out to be superb mothers. I think early interventions that have the effect of stabilizing the environment and supporting the mothers are important. For the record, we have tried giving these monkeys the SSRI drugs fluoxotine and sertraline. When we gave these drugs to aggressive and impulsive young adult monkeys we observed a great improvement in their behaviour, but as soon as we took them off the drugs they reverted to their previous patterns, so there apparently was no long-term prophylactic influence of these drugs.

Hinde: All your monkeys are provisioned. In bad circumstances these aggressive monkeys may be better off.

Suomi: There is a case that is relevant to this. The researchers who have been studying rhesus monkeys on the island of Cayo Santiago had an unfortunate situation where the monkeys basically ran out of food and water for over a month several years ago. The researchers have good records of who survived and who didn't. It was a brutal situation and a lot of monkeys didn't make it. Someone should be able to get the information on the allelic status of these monkeys and look at this question.

Nelson: Are there any endocrine correlates with the CSF 5-HIAA levels?

Suomi: I wish there were. We have been looking for endocrine or other peripheral 'proxies' for CSF 5-HIAA concentrations for a while. We thought we had found one in a measure of salivary prolactin, but it turns out that the assays don't give any reliable results for about a third of the samples. We are really looking for a peripheral silver bullet, because if this set of findings carried over to humans, one could start looking at these issues directly in humans. It will be a long time before any basic researcher is allowed to take CSF samples from normal human populations.

Nelson: Most of your comparisons were between peer-reared and mother-reared monkeys, and you used the description 'good' mothers. Do you have any data about the differences?

Suomi: The subjects in the studies were indeed good mothers; they were in our breeding colony. Within our overall colony there is a substantial range of maternal capabilities. Indeed, there is a nice study of a family of rhesus monkeys at the Yerkes Primate Center's field station who generation after generation have been abusive to their offspring. Preliminary analyses of these data have shown that these females display alterations of HPA axis activity and circadian rhythmicity. Within our own monkey colony, some females are clearly less competent maternally than others, especially among those who were peer-reared as infants. In wild populations, approximately 10% of primiparous females lose their offspring because of neglect and abuse. They have better success with subsequent offspring. We are now starting to look within our colony at the range of parenting of mothers, first to determine whether we can find the same kind of consequences in terms of the behaviour of the offspring, and second whether there is anything unusual genetically about these offspring. Our preliminary analyses suggest that the patterns that we are observing in the offspring of the least competent mothers parallels the patterns we see with the peer-reared monkeys.

Keverne: The differences you got in 5-HIAA with age were remarkable. There was a 100% change between day 14 and day 150, and a further 50% between one year and three years. What is going on in the brain to bring about these massive changes?

Suomi: I am not the person to ask about any neurochemical mechanisms that might underlie these dramatic developmental declines in CSF 5-HIAA concentrations. But we have data showing that these concentrations decline in *all* subjects during development, but the rearing condition differences essentially remain comparable the same at each age examined. The more interesting observation, as you pointed out, is this normative drop that is absolutely enormous, but it is species-normative.

References

Kaplan JR, Rontenot MB, Berard J, Manuck SB, Mann JJ 1995 Delayed dispersal and elevated monoaminergic activity in free-ranging rhesus monkeys. Am J Primatol 26:431–443

Mehlman PT, Higley JD, Faucher I et al 1995 Correlation of CSF 5-HIAA concentration with sociality and the timing of emigration in free-ranging primates. Am J Psychiatry 152:907–913

Trefilov A, Berard J, Drawczak M, Schmidtke J 2000 Natal dispersal in rhesus macaques is related to serotonin transporter gene variation. Behav Genet 3:295–301

The role of monoamine oxidase A, *MAOA*, in the aetiology of antisocial behaviour: the importance of gene–environment interactions

Ian W. Craig

Social, Genetic, and Developmental Psychiatry Centre, Institute of Psychiatry, King's College London, PO82, De Crespigny Park, London SE5 8AF, UK

Abstract. Reports from both human studies and animal models suggest that *MAOA* may have a key role in aggression. Differences in the copy number of a repeat motif in the promoter of *MAOA* appear to regulate its activity. We review the evidence that suggests activity levels of this enzyme may play a key role in modulating antisocial/aggressive behavioural outcomes. Two common alleles in the human population have been identified which confer either high or low transcriptional functionality. The gene is X-linked and males can, therefore, be typed as 'low' or 'high' types. A key feature that has emerged from both our own studies and replicated recently by others, is that high activity variants of the gene appear to confer a protective influence against maltreatment, such that maltreated males with a high-*MAOA* activity genotype were less likely to develop antisocial problems, an observation that may explain part of the variability in developmental outcomes associated with maltreatment.

2005 Molecular mechanisms influencing aggressive behaviours. Wiley, Chichester (Novartis Foundation Symposium 268) p 227–241

A role for genetic factors in aggression?

It is our common experience that there is a strong sex bias in antisocial behaviour (ASB) in humans, with males having much greater tendency towards violence than females. For example, in a longitudinal study of about 1000 individuals of both sexes from an age of three to 21, reviewed in depth by Moffit et al (2001), ASB was found to be overall 2.4 times more prevalent in males than females. Evidence also points to a higher heritability of aggression in males, whereas common environment may be more important in females (Miles & Carey 1997, Vierikko et al 2003).

Although there is a correlation between a rise in hormone levels and ASB in adolescent males and the relationship between testosterone and aggression has been demonstrated through correlational and experimental studies in animals (Turner 1994), evidence for a direct role of testosterone in determining aggression in humans is not clear-cut (e.g. see Archer 1991, Turner 1994, van Honk et al 1999). We must therefore have an open mind concerning the possible influence of other genes that may contribute to a greater propensity to aggression in males.

This chapter reviews our work and that of others in examining the potential role of specific genetic factors, unrelated to those directly implicated in sex determination, in predisposing this sex imbalance of behavioural phenotypes. Given the observation that experience of abuse greatly increases the probability of antisocial personality and subsequent violent offending in males (e.g. Rutter et al 1998), it focuses on the role of *MAOA* variants in ASB and the cycle of violence in maltreated males.

MAOA as a candidate gene for behavioural disorders

The role of MAOA in metabolism of neurotransmitters

The main role for *MAOA* is thought to be in degrading serotonin following its re-uptake from the synaptic cleft; although it is also capable of degrading both noradrenaline and dopamine. It is therefore plays a key role in the regulation of synaptic activity and alterations in its activity produced by pharmacological intervention or through genetic variants are likely to have profound effects on behaviour. Indeed, drugs inhibiting its activity have long been employed in the treatment of behavioural disorders — particularly depression.

Variants associated with altered expression/activity of MAOA

Both restriction fragment length polymorphisms (RFLPs) and a variety of microsatellite markers have been detected around the *MAOA* loci including *Eco*RV and *Fnu*4HI polymorphisms (Hotamisligil & Breakfield 1991), a variable number tandem repeat (VNTR) in intron 1 (Hinds et al 1992) and an (AC)n microsatellite in intron 2 of the *MAOA* gene (Black et al 1991), all of which have been employed in association studies with a range of behavioural disorders (see Fig. 1). Following the identification of the promoter region of *MAOA* by Denney et al (1994), a major development came through the report of a novel motif localized 1.2 kb upstream from the *MAOA* coding region, comprising a 30 bp repeat existing in 3, 3.5, 4 or 5 copies (Sabol et al 1998). Importantly, they demonstrated that the copy number had a significant effect on the level of transcription achieved when the motifs were coupled to a reporter gene and transfected into a variety of cell types. As in all subsequent publications, the two major alleles were found to be those with three or four repeats of the 30 bp motif,

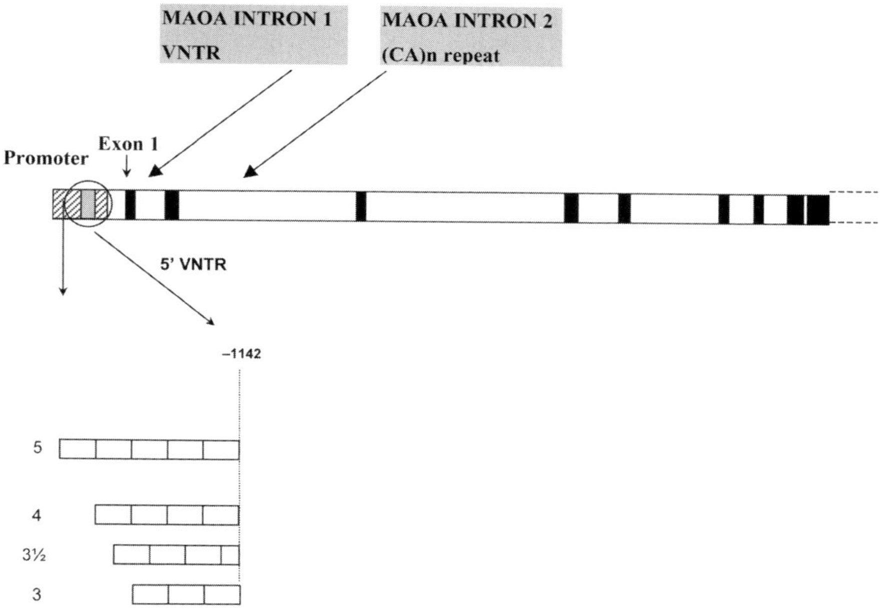

FIG. 1. Diagram of upstream VNTR and locations of intron microsatellite and VNTR.

with the lower copy number showing reduced transcriptional activity and the current general consensus is that longer repeat number alleles have higher transcriptional activity than those with shorter repeats (Deckert et al 1999, Denney et al 1999). Sabol et al (1998) also reported significant variations in the distribution of the uVNTR alleles among major population groups, suggesting that population stratification could be a major confound in the interpretations of association studies. Deckert et al (1999) in addition to confirming the importance of promoter variants in regulating *MAOA* expression by transfection studies demonstrated that the high activity variant(s) were associated with panic disorder in females (see below). It is probable that associations based on the RFLPs and microsatellites reflect their allelic association with the uVNTR.

MAOA variants and antisocial/aggressive behaviours

A watershed for investigations into the potential role for *MAOA* variants in the aetiology of ASB and conduct disorders at the population level was the report of Brunner et al (1993). They described the segregation in males from a Dutch pedigree of a complex behavioural syndrome (including impulsive aggression) and a nonsense mutation resulting in zero activity of the enzyme. Apart from some males with deletions covering the Norrie gene and the adjacent monoamine oxidase A and B loci, to date this has been the only report of the phenotypic consequences of a null mutation (see Shih & Thompson 1999). The patients with Norrie disease show profound and complex behaviours, apart from blindness and in some cases sensorineural deafness, which makes any assessment of the contribution of the gene deletions to possible impulsive aggression difficult. Since this time, while there is undoubtedly a bias in the reporting of positive findings, there have been several reports suggesting a role for variants at the *MAOA* locus in the context of aggression and/or ASB.

Vanyukov et al (1995) first carried out an investigation of a possible association between the intron 2 microsatellite with aggression in adolescents and although found no significant correlation between aggression scales and repeat length, they did find a trend towards association between longer microsatellite alleles and a categorical diagnosis of conduct disorder. Lawson et al (2003) examined the hypothesis that male attention deficit hyperactivity disorder (ADHD) probands with comorbid conduct disorder may have reduced *MAOA* activity and provided evidence in case control studies for an excess of the low activity uVNTR allele (three copies of the repeat) in individuals with this combined phenotype ($P=0.025$). They also showed a strong trend to excess transmission of this low activity allele in TDT ($P=0.08$). Saito et al (2002) examined the potential association of the uVNTR in Finnish males with alcoholism and impulsive behaviour. Both type 1 and type 2 alcoholics were investigated (the latter characterized by antisocial behaviour). Although no strong effect was detected, there was an overall trend for an association between the low activity variant and alcoholism. Schmidt et al (2000) also noted an excess of the low activity allele in antisocial alcoholics compared with those exhibiting anxious-depressive symptoms (51%:23%). A similar conclusion was reached by Samochowiec et al (1999) who investigated the uVNTR and alcoholism/antisocial personality disorder in a large cohort of German descent. The frequency of the low activity (three-repeat) allele was significantly increased in the 59 antisocial alcoholics compared to the 185 control subjects ($P=0.008$); but no significant difference in allele frequencies between the non-antisocial alcoholics and controls. Hsu et al (1996) also found evidence linking *MAOA* variants (both *Eco*RV and *Fnu*4HI RFLPs and the intron 2 microsatellite polymorphisms) with susceptibility to alcoholism among Han Chinese. Other studies, however, have been unable to replicate this observation (Lu et al 2002). Similarly, Parsian et al (2003) and Koller et al (2003) failed to detect a significant association between type II alcoholics and variants of the uVNTR in similar ethnic groups to those in which positive findings had been reported previously. Furthermore, Manuck et al (2000) found individuals with the low-activity allele showed less dispositional aggressiveness and impulsivity ($P < 0.015$) in an investigation of inter-individual variability in aggressiveness, impulsiveness and CNS serotinergic responsiveness on 110 males. Nevertheless, overall, there does seem to be evidence linking low expression with increased tendency to ASB.

A role for *MAOA* deficiency in promoting aggression is further supported by studies on transgenic mice having a 17 kb deletion embracing exons 2 and 3 of the *Maoa* gene which abolished enzyme activity in brain and liver (Cases et al 1995). Behavioural studies on the adult males indicated heightened aggression in resident–intruder tests and also increased inappropriate grasping during courtship.

Most recently, it has been reported that a similar promoter variant system exists in primates. In macaques, where the repeat motif is 18 base pairs, Lesch and colleagues have shown that it too can regulate the expression of *MAOA* and fundamental to the potential role of *MAOA* in human ASB, appeared to influence aggression levels in competing for food (T. K. Newman and K. P. Lesch, personal communication).

There is, therefore, an accumulation of evidence suggesting association of *MAOA* markers tracking low activity variants to be associated with externalizing symptoms, particularly ASB in males. In contrast, a number of reports suggest that high activity alleles may be associated with internalizing behaviours including anxiety, neuroticism and panic disorder (e.g. Karayiorgou et al 1999, Deckert et al 1999, Schulze et al 2000, Eley et al 2003). As with many candidate gene studies, there is frequently a failure to replicate some of the original suggestive data. This is, of course, to be expected given the heterogeneity in the studies with respect to diagnosis and ethnicity. Similarly, failure to replicate is often a consequence of insufficient power. In addition, one of the main confounds of such investigations is the difficulty in controlling for the environmental effects. It is for these reasons that we sought to investigate the role of the *MAOA* activity variants in the aetiology of ASB in males in a manner that allowed for the control of significant environmental variants by investigating variability in antisocial outcomes in male children who are maltreated (Caspi et al 2002). This longitudinal investigation of a large representative sample of the general population takes full advantage of the wealth of data available concerning both history of environmental adversity and measures of antisocial behaviour.

It was first established that there was no significant association of *MAOA* genotype with maltreatment category assignment. Independent source information was used to evaluate convictions for violent crimes, a personal disposition toward violence, symptoms of antisocial personality disorder at age 26 and adolescent conduct disorder (assessed according to DSM-IV criteria). Analysis of common factor loadings showed that all four measures index liability to ASB. Whichever of these measures is examined, we have shown that for males

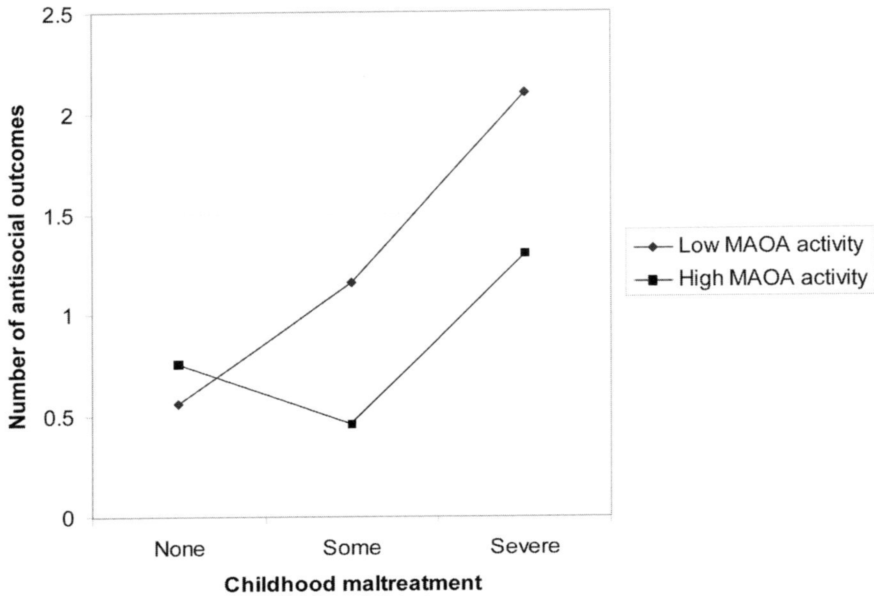

FIG. 2. Antisocial outcomes as a consequence of childhood maltreatment by MAOA genotype in males.

there is an association between maltreatment and ASB conditional on the individual's *MAOA* genotype with *P* values of 0.05, 0.10, 0.04 and 0.06, respectively. In contrast, for males with high activity alleles, maltreatment did not confer a significant risk to be convicted of violent crime or to exhibit other ASB. Figure 2 illustrates the overall increasing tendency to antisocial acts with more extreme childhood maltreatment for males with the low activity alleles, but not for those 'protected' by the high activity alleles.

Given that the opportunity both to assess adverse life events and to obtain in depth information on ASB in such detail is rare, it is remarkable that replication of the observations has already been achieved (Foley et al 2004). This study was based on individuals with relevant data from 514 white male twins from the community-based longitudinal Virginia Twin Study for Adolescent Behavioral Development. Individuals were between 8–17 at the time of entry into the study. The measures for childhood maltreatment differed from those employed by Caspi et al (2002) in that adversity was measured by exposure to three known risk factors; inter parental violence, parental neglect and inconsistent discipline. Nevertheless, employing DSM-III-R criteria using the Child and Adolescent Psychiatric Assessment

(CAPA)-Child and Parent version on four occasions, the authors reported a highly significant increase in conduct disorder with increased exposure to childhood adversities for males with low activity *MAOA* genotype, a consequence which those with high activity alleles were protected against (Fig. 3).

The impact of stress on neurotransmitter pathways

Even though abuse in childhood increases the risk of later criminality by about 50%, most maltreated children do not later exhibit ASB. We have hypothesized that the vulnerability of those that do respond in this way could be conditional upon various genetic factors such as the variability in the uVNTR of the *MAOA* locus (Caspi et al 2002). An important question that arises is therefore, what is the link between exposure to stressful events and the monoamine neurotransmitter pathways? A range of animal studies indicates such features as early maternal deprivation may affect the noradrenaline, serotonin and dopamine systems with repercussions through to adulthood (reviewed in Caspi et al 2002). Evidence that stress in social encounters and being the recipient of defeats in aggressive confrontations may have a direct effect on *MAOA* levels in the brain comes from studies on CBA/Lac male mice. Filipenko et al (2002) monitored expression of *MAOA* and *SERT* by multiplex RT-PCR and found both were significantly elevated. This may reflect a compensatory response to activation of the serotonergic system functioning activated by social stress. Could the connection between the apparent risk conferred by the low activity allele reflect a relative inability to respond to the effects of stress arising from maltreatment? An inability to increase *MAOA* levels that could normally cushion the chronic release of excess neurotransmitters might be displaced into violent outbursts and other aspects of ASB.

Evolutionary significance

Gilad et al (2002) have carried out an in depth study of linkage disequilibrium (LD) across five regions covering 90 kb of the *MAOA* locus in 56 males from seven different ethnographic groups. LD appeared to be higher than expected for typical nuclear loci under panmictic models, suggesting a pattern accentuated by positive selection, potentially acting on *MAOA*-related phenotypes. Such observations could be interpreted to suggest that selection for hyperactivity and aggression in males at these two loci may have played a part in the ancestry of human populations. Coupled with the observations of Lesch and Newman (see above), it appears that uVNTR *MAOA* variants may have played important selective roles in the recent ancestry of humans and that the consequences of this reverberate in contemporary society.

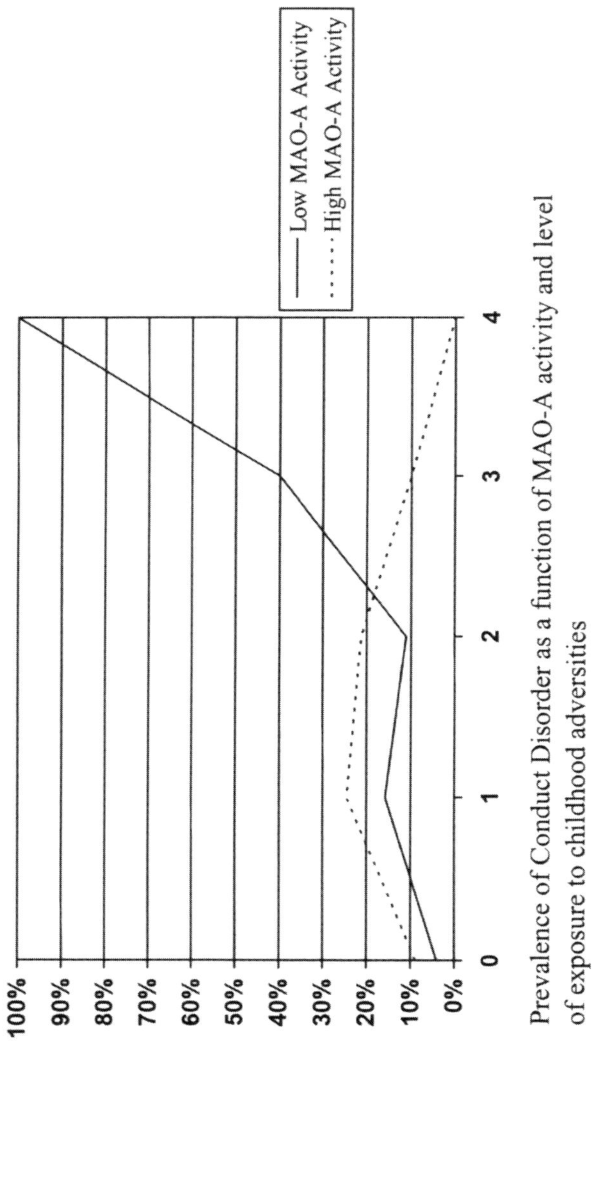

FIG. 3. Prevalence of Conduct Disorder as a function of MAOA activity and level of exposure to childhood adversities. From Foley et al (2004) with permission.

Conclusions

Several studies have independently indicated that the shorter of the two prevalent variants located in the promoter of MAOA results in lower level of gene expression. Critically, the strongest evidence for a role for *MAOA* activity variants in ASB emerges when gene×environment interactions are considered and the role for the low activity promoter in the cycle of violence first suggested by Caspi et al (2002) has now been further supported by the studies of Foley et al (2004).

Finally, the recent reports that similar polymorphic activity variants have been discovered in the brains of primates suggests both their considerable evolutionary importance and that they may be maintained by some form of balancing, or frequency-dependent, selection.

Acknowledgements

It has been a privilege to have been able to work with Avshalom Caspi and Terrie Moffit on this topic and I am profoundly grateful to all participants in the Dunedin programme.

References

Archer J 1991 The influence of testosterone on human aggression. Br J Psychol 82:1–28

Black GC, Chen ZY, Craig IW, Powell JF 1991 Dinucleotide repeat polymorphism at the MAOA locus. Nucleic Acid Res 19:689

Brunner HG, Nelen M, Breakefield XO, Ropers HH, van Oost BA 1993 Abnormal behavior associated with a point mutation in the structural gene for monoamine oxidase A. Science 262:578–580

Cases O, Seif I, Grimsby J et al 1995 Aggressive behaviour and altered amounts of brain serotonin and norepinephrine in mice lacking MAOA. Science 268:1763–1766

Caspi A, McClay J, Moffitt TE et al 2002 Role of genotype in the cycle of violence in maltreated children. Science 297:851–854

Deckert J, Catalano M, Syagailo YV et al 1999 Excess of high activity monoamine oxidase A gene promoter alleles in female patients with panic disorder. Hum Mol Genet 8:621–624

Denney RM, Sharma A, Dave SK, Waguespack A 1994 A new look at the promoter of the human monoamine oxidase A gene: mapping transcription initiation sites and capacity to drive luciferase expression. J Neurochem 63:843–856

Denney RM, Koch H, Craig IW 1999 Association between monoamine oxidase A activity in human male skin fibroblasts and genotype of the MAOA promoter-associated variable number tandem repeat. Hum Genet 105:542–551

Eley TC, Tahir E, Angleitner A et al 2003 Association analysis of MAOA and COMT with neuroticism assessed by peers. Am J Med Genet 120B:90–96

Filipenko ML, Beilina AG, Alekseyenko OV, Dolgov VV, Kudryavtseva NN 2002 Repeated experience of social defeats increases serotonin transporter and monoamine oxidase A mRNA levels in raphe nuclei of male mice. Neurosci Lett 321:25–28

Foley DL, Eaves LJ, Wormley B et al 2004 Childhood adversity, monoamine oxidase A genotype, and risk for conduct disorder. Arch Gen Psych 61:738–744

Gilad Y, Rosenberg S, Przeworski M, Lanvet D, Skrecki K 2002 Evidence for positive selection and population structure at the human MAOA gene. Proc Natl Acad Sci USA 99: 862–867

Hinds HL, Hendriks RW, Craig IW, Chen ZY 1992 Characterization of a highly polymorphic region near the first exon of the human MAOA gene containing a GT dinucleotide and a novel VNTR motif. Genomics 13:896–897

Hotamisligil GS, Breakefield XO 1991 Human monoamine oxidase A gene determines levels of enzyme activity. Am J Hum Genet 49:383–392

Hsu YP, Loh EW, Chen WJ, Chen CC, Yu JM, Cheng AT 1996 Association of monoamine oxidase A alleles with alcoholism among male Chinese in Taiwan. Am J Psychiatry 153:1209–1211

Karayiorgou M, Sobin C, Blundell ML et al 1999 Family-based association studies support a sexually dimorphic effect of COMT and MAOA on genetic susceptibility to obsessive-compulsive disorder. Biol Psychiatry 45:1178–1189

Koller G, Bondy B, Preuss UW, Bottlender M, Soyka M 2003 No association between a polymorphism in the promoter region of the MAOA gene with antisocial personality traits in alcoholics. Alcohol Alcoholism 38:31–34

Lawson DC, Turic D, Langley K et al 2003 Association analysis of monoamine oxidase A and attention deficit hyperactivity disorder. Am J Med Genet 116B:84–89

Lu RB, Lee JF, Ko HC, Lin WW, Chen K, Shih JC 2002 No association of the MAOA gene with alcoholism among Han Chinese males in Taiwan. Prog Neuropsychopharmacol Biol Psychiatry 26:457–461

Manuck SB, Flory JD, Ferrell RE, Mann JJ, Muldoon MF 2000 A regulatory polymorphism of the monoamine oxidase-A gene may be associated with variability in aggression, impulsivity, and central nervous system serotonergic responsivity. Psychiatry Res 95:9–23

Miles DR, Carey G 1997 Genetic and environmental architecture of human aggression. J Pers Soc Psychol 72:207–217

Moffitt TE, Caspi A, Rutter M et al 2001 Sex differences in antisocial behaviour: conduct disorder, delinquency and violence in the Dunedin Longitudinal Study. Cambridge University Press, Cambridge

Parsian A, Cloninger CR, Sinha R, Zhang Z-H 2003 Functional variation of monoamine oxidase A and subtypes of alcoholism: haplotype analysis. Am J Med Gen Part B 117B:46–50

Rutter M, Giller H, Hagell 1998 Antisocial behaviour by young people. Cambridge University Press, Cambridge

Sabol SZ, Hu S, Hamer D 1998 A functional polymorphism in the monoamine oxidase A gene promoter. Hum Genet 103:273–279

Saito T, Lachman HM, Diaz L et al 2002 Analysis of monoamine oxidase A (MAOA) promoter polymorphism in Finnish male alcoholics. Psychiatry Research 109:113–119

Samochowiec J, Lesch KP, Rottmann M et al 1999 Association of a regulatory polymorphism in the promoter region of the monoamine oxidase A gene with antisocial alcoholism. Psychiatry Res 86:67–72

Schmidt LG, Sander T, Kuhn S et al 2000 Different allele distribution of a regulatory MAOA gene promoter polymorphism in antisocial and anxious-depressive alcoholics. J Neural Transm 107:681–689

Schulze TG, Muller DJ, Krauss H et al 2000 Association between a functional polymorphism in the monoamine oxidase A gene promoter and major depressive disorder. Am J Med Genet 96:801–803

Shih JC, Thompson RF 1999 Monoamine oxidase in neuropsychiatry and behaviour. Am J Hum Genet 65:593–598

Turner AK 1994 Genetic and hormonal influence on male violence. In: Archer J (eds) Male Violence. Routledge, New York

Van Honk J, Tuiten A, Verbaten R et al 1999 Correlations among salivary testosterone, mood, and selective attention to threat in humans. Horm Beh 36:17–24
Vanyukov MM, Moss HB, Yu LM, Deka R 1995 A dinucleotide repeat polymorphism at the gene for monoamine oxidase A and measures of aggressiveness. Psychiatry Res 59:35–41
Vierikko E, Pulkkinen L, Kaprio J et al 2003 Sex Differences in genetic and Environmental Effects on Aggression. Aggress Behav 29:55–68

DISCUSSION

Brodkin: In your data, high MAOA activity was protective in the face of severe maltreatment, but the severe maltreatment and high activity MAOA subgroup seems to have increased variability in some of the measures of violence and antisocial problems. One way to view this might be that within this subgroup, the high MAOA activity may be sufficient to be protective for some individuals, but not for others. The variability within this subgroup may come from the effect of other gene variants. Might a future strategy might be to look not only at gene–environment interactions, but also gene–gene–environment or gene–gene–gene–environment interactions? How hard is this to do?

Craig: Practically it is easy to do. The interpretation is much more difficult, and the more variables there are the more tests required and this removes a lot of significance as a result of multiple testing. The way forward is to look at precisely this kind of thing in a stepwise manner. For example, you could take the genes in the serotonin system and work out a hypothesis based on what you see there, and then go on to test this in a different system.

Brodkin: Hypothetically, if someone in the future wanted to have their genome scanned to assess their risk for aggressive behaviours, they would probably have to look across many different genes and consider gene–gene interactions, as well as gene–environment interactions.

Manuck: I enjoyed your presentation very much, and the whole Dunedin study is spectacular — really, one of a kind. The article describing interactive effects of MAOA variation and childhood maltreatment on later violence and antisocial behaviour will surely prove a seminal publication in this area (Caspi et al 2002). As we discussed yesterday, there are nonetheless some discrepancies in the MAOA literature, so it might be useful to consider a few of these and how they might be reconciled. Among the early MAOA studies was one of our own (Manuck et al 2000), in which we observed that variation in impulsive aggression (as seen in our non-patient sample, and defined by psychometric reduction of multiple instruments) was greatest among men having alleles associated with higher rates of *MAOA* gene transcription. Although not a large study, we also found that men with a high transcription allele showed reduced brain serotonergic responsivity by standard fenfluramine challenge, and that adjusting for variability in serotonergic response eroded the statistical effect of *MAOA* genotype on participants' aggression scores to nonsignificance (Manuck et al 2002). The latter finding suggests that this *MAOA* promoter polymorphism influences an aggressive phenotype via allele-specific variation in CNS serotonergic function. However, the direction of association we observed (viz., greater aggressiveness in men with allelic variation predicting high MAOA activity) is apparently opposite the observations you and your colleagues found among individuals who experienced childhood abuse in the Dunedin cohort.

The very recent report of an *MAOA* × environment interaction paralleling the Dunedin finding, as presented by an independent group of investigators, would seem to swing the balance toward an association of aggressiveness with the low activity *MAOA* genotype (Foley et al 2004). Yet I believe a close reading of this paper suggests a more complex picture that is worth thinking about briefly. In the Foley et al (2004) study, it is reported that children and adolescents carrying low activity *MAOA* alleles were at increased risk of conduct disorder only if they had experienced an adverse rearing environment characterized by inconsistent discipline and parental neglect and conflict. So far, so good. In their full logistic regression model, however, the authors acknowledge that on controlling for early adversity and the interaction of adversity and *MAOA* genotype, low MAOA activity was associated with lower (not higher) risk of conduct disorder across their overall sample of 514 boys (conduct disorder was present in 8.6% and 12.7% of participants having low and high MAOA activity genotypes, respectively). The interaction that suggests replication of the Dunedin study derives from a very high prevalence of conduct disorder among the few boys who were in the highest levels of childhood adversity and who also happened to have low *MAOA* activity alleles. But there were only three such boys (accounting for 5% of the 59 cases of conduct disorder in the full sample). Also, the two highest strata of adversity that drive the statistical interaction encompassed less than 4% of the study cohort. Thus, one interpretation of this study's findings would be that the high MAOA activity genotype is generally associated with an increased risk of conduct disorder (see also Beitchman et al 2004), but that low activity alleles confer markedly increased risk in the context of infrequently encountered, yet severely adverse, rearing environments (as in your study). Is it conceivable that so complex a relationship may exist among developmental experiences, common genetic variation, and aggressive phenotypes? I wonder, too, if perhaps the nature of the conduct problems seen among those reared under egregious adversity differ in significant ways from the aggressiveness of individuals exposed to less extreme developmental environments.

Craig: That is a reasonable interpretation of the Dunedin study, because the lines do cross over; but not significantly, at the low adversity end. I was interested to see how we could divide up the personality measures to see whether there are differences in the way in which the response goes. I think what you are saying is that maybe there is.

Manuck: It might be interesting to conduct an item analysis with respect to components of antisocial behaviour measured in the Dunedin study. This might be a way of determining whether aspects the aggressive phenotype associated with childhood adversity and the low activity *MAOA* genotype differ from aggressive behaviour seen in participants of other genetic and environmental backgrounds.

Suomi: The aggression data you showed were not the same sort of impulsive aggressive behaviours that I was describing in my primate studies. This goes right back to the heart of the first paper at this meeting, on characterizing the different types of aggression. Rhesus monkeys are notoriously aggressive as a species, but much of that aggression serves positive social purposes. Those forms have to do with enforcing dominance hierarchies, and they largely involve threats and barks as opposed to actual physical attacks. The aggression measured in that paper that also showed an interaction was the so-called socially appropriate or dominance-related aggression. These forms are different in both the pattern of this behaviour and its hormonal correlates from the impulsive aggression I was talking about in my presentation. The latter has an inverse correlation with 5-HIAA concentrations and no relationship with testosterone levels, whereas in males the type of aggression that was depicted in the data from Tim's study is unrelated to 5-HIAA concentrations and is positively correlated with testosterone levels in the subjects involved.

C Blanchard: The last two papers have raised an interesting question. In your group you were talking about bringing people back from all over the place. These people had evidently moved. Was there any suggestion that people with certain characteristic left quicker, or went to strange places?

Craig: The numbers might be a bit small to get anything out of it. But the idea that the explorer or entrepreneur has characteristics that makes them colonize new territories is a nice idea.

Keverne: You showed us nice data on MAOA and 5-HT. Does the low activity versus high activity allele have any differential effects on the other catecholamines?

Craig: Not directly. Imaging studies are currently being done.

Lesch: From knockout mice we know that MAOA in the brain is primarily modulating 5-HT and has much less impact on dopamine and noradrenaline. This is unlike MAOB.

Martinez: How did you measure child maltreatment? Did you distinguish between abuse by parents, close relatives or strangers? Did you categorize the severity?

Craig: No, it was an all-or-nothing observation. Harsh discipline was a parental report which will always be biased in some kind of way, because people don't want to admit to what they are doing. Most of it is measured retrospectively by the individual. If they remember it, it must have been serious enough for them to recall.

Martinez: Did you separate the subtypes?

Craig: They were separated in the original reports, since then they have been clumped together.

Martinez: I am not an expert in child conduct but I would imagine that the behaviour of the parents is also influenced by the behaviour of the children.

Craig: We have looked at the allele frequencies of the individuals who were given harsh, intermediate or low maltreatment, and there was no difference in the allele frequencies. So there is no sectoring of evoking parental responses on the basis of the *MAOA* genotype.

Pfaff: From the Gilad et al (2002) paper you cited, can you give us an idea of the nature of the argument for positive selection? In particular, what are the alternatives?

Craig: Recombination is normally going to separate out the different alleles. If you find particular combinations of alleles persisting at higher frequency than you would anticipate on the basis of their individual frequencies, it means that recombination hasn't happened. We know roughly over the genome the way that recombination is separating out alleles. If you find a higher-than-expected combination of specific alleles at a particular site, this suggests that something to counter the anticipated recombination must have happened. There are various explanations for how that could be. One is selection, which says that this is a good combination of alleles, which may be in parallel with a functional variant that is doing something useful. The other is a founder effect or a migration; anything which in a population is going to capture a particular haplotype.

Pfaff: Or something to with the chemistry of the recombinase.

Craig: It could also be recombination differentials across different parts of the genome. The linkage disequilibrium map is quite heterogeneous. It does look quite interesting.

Manuck: Is any of the *MAOB* variation that you identified functional?

Craig: No. We have this VNTR at the 5′ end of the gene. It is about 3 kb away, but I don't think it is functional.

References

Beitchman JH, Mik HM, Ehtesham S, Douglas L, Kennedy JL 2004 MAOA and persistent, pervasive childhood aggression. Mol Psychiatry 9:546–547

Caspi S, McClay J, Moffitt TE et al 2002 Role of genotype in the cycle of violence in maltreated children. Science 297:851–854

Foley DL, Eaves LJ, Wormley B et al 2004 Childhood adversity, monoamine oxidase A genotype, and risk for conduct disorder. Arch Gen Psychiatry 61:738–744

Gilad Y, Rosenberg S, Przeworski M, Lanvet D, Skrecki K 2002 Evidence for positive selection and population structure at the human MAOA gene. Proc Natl Acad Sci USA 99:862–867

Manuck SB, Flory JD, Ferrell RE, Mann JJ, Muldoon MF 2000 A regulatory polymorphism of the monoamine oxidase-A gene may be associated with variability in aggression, impulsivity, and central nervous system serotonergic responsivity. Psychiatry Res 29: 651–668

Manuck SB, Flory JD, Muldoon MF, Ferrell RE 2002 Central nervous system serotonergic responsivity and aggressive disposition in men. Physiol Behav 77:705–709

Final general discussion

Pfaff: I'd like to ask Caroline and Bob Blanchard to begin this discussion by raising issues from this meeting they would like us to revisit.

C Blanchard: I would like to start by bringing up some information on neural systems involved in defensive behaviours, potentially including defensive aggression. These are based on the work of Newton Canteras (Canteras et al 2001), analysing Fos expression after a rat has been exposed to a cat, and on retrograde and anterograde tracing from some of the sites showing such Fos enhancement. Newton has focused on a medial hypothalamic zone including the anterior hypothalamic nucleus, the dorsomedial part of the ventromedial nucleus, and the dorsal premammillary nucleus. This system receives input from specific portions of the amygdala, and the prefrontal cortex, as well as the lateral septal nucleus, and the interfascicular nucleus of the bed nucleus of the stria terminalis. These Fos studies don't necessarily mean that these structures are functional in defence, but we have been working with Newton in making lesions in some of them to see if they alter defensive behaviours of rats to a cat. The most dramatic results are that both ibotenic and electrolytic lesions of the dorsal premammillary nucleus substantially reduce a number of different measures of defensiveness to predator stimuli, but damage to some other components of the system, notably the lateral amygdala, and the ventral but not dorsal hippocampus, also reduce some aspects of defence.

There is also some basis for comparison of this work with what happens in resident–intruder and offensive aggression. Animals in resident–intruder confrontations (N. Canteras, personal communication) also show enhanced Fos activity, but in the medial amygdala this involves a different area than when the rat is exposed to a cat. Specifically, for cat exposure it was the posterior ventral part of the medial amygdala, whereas in the resident–intruder test, both participants—resident and intruder—showed Fos expression in the posterior dorsal part of the medial nucleus. Residents and intruders were, however, differentiated by the amount of signal in that particular area, with the residents showing heavier signal than the intruders. If we move back into the hypothalamus, Fos is heavily expressed in the dorsal premammillary nucleus, but not in the ventral premammillary nucleus during cat exposure. However, in the resident–intruder test both participants showed little Fos activity in the dorsal premammillary nucleus and considerable activity in the ventral part. Again, this was quantitatively different for the resident and the intruder.

Finally, complications abound in these data and their interpretations. If rats with dorsal premammillary nucleus lesions that show substantial reductions in freezing to cat odour are given a post-shock freezing test, they freeze normally. This is not what you would expect if this area is part of a general motivational system underlying defensiveness, or if it is part of a system directly involved with freezing. Also, if you put such animals in a test of response to female odour, there is no significant difference between the lesioned and sham-operated group, so this is not a purely odour-based or pheromonal odour-based system either.

R Blanchard: The central issue is that there are around 30 labs working on the neural systems involved in defence and defensive aggression. These systems are beginning to be elucidated. But we really have no idea of the extended neural systems that modulate offensive behaviour. This remains an important problem for those of us who want to understand the biology of the system, including the genetics.

Olivier: In my PhD research I was burning holes in the hypothalamus and looking for resident–intruder aggression. One of the most prominent areas for evoking offensive aggression was the ventral premammillary nucleus. The dorsal part may be involved in defence and the ventral part in offence. Is this a possibility?

C Blanchard: Yes, that is one interesting possibility. However, the data indicate that in a resident–intruder test, both the resident and the intruder are showing Fos expression in the ventral premammillary nucleus, but the intruder shows heavier Fos activity than the resident in this area.

Olivier: Together with the increased resident–intruder aggression, which went up from 20% to almost 80% of the time, there were also more special behaviours. It wasn't a pure aggression effect.

R Blanchard: Indeed, the ventral premammillary nucleus may be central in the regulation of responses to pheromonal odours related to sex. It may be very closely related to both sex and aggression.

Pfaff: Another topic that might be worth discussing is that I noticed that Jan Koolhas' group published a Society for Neuroscience abstract in 2002 which was very provocative. Just briefly, I'll mention the results. He has rats with interesting polymorphisms on the Y chromosome. He did a rather sophisticated factor analysis of the various indices of aggression. He came out with a factor that didn't explain much of the variance but which was associated with wildly aggressive behaviour that reminded him of a concept similar to 'violence'. Another question that might be good for general discussion is further consideration of the roles of the quantitative trait loci (QTLs) on the X chromosome.

Brembs: One question I think we should consider is this: what is pathology in aggressive behaviour? What I mean by this question is that if we define pathology, is this just one end of a Gaussian distribution when we look at a histogram, or is it discontinuous? Are there criteria that are not continuous?

Martinez: I can give an example with battering men. Most people might think that men who batter their wives have a mental illness. This isn't true. It is considered that only 25% have some antisocial personality disorder. The other 75% are lovely people in other contexts but aggressive at home. In many cases battering is related mostly to cultural factors rather than to mental illness.

R Blanchard: Wouldn't the fact that they are battering be enough to define this as a pathology? The fact that it doesn't correlate with mental illness doesn't mean that it can't be characterized as a pathology.

Suomi: What if it is a cultural norm? These distinctions are relative.

C Blanchard: It goes back to the functional aspects of aggression. Most guys who beat their wives get some sort of reinforcement for it. I once interviewed a wife beater and asked him what happened as a result of the beating. He said he wins the argument.

Craig: Going back to what we were talking about earlier, the effect which we are seeing with regard to upbringing is in a very small number of individuals who are very disturbed and antisocial. Few individuals carry out most of the misdemeanours. They have the pathology. I don't know how they fit into a continuum.

Manuck: In psychiatry, there is currently much debate regarding the extent to which clinically recognized psychopathologies comprise categorical entities (analogous to diseases of discrete aetiology) or the extremes of otherwise commonly distributed traits of personality or temperament (liability dimensions). I am not conversant in criminology, but there is one obvious sense in which criminality would seem a categorical construct. That is, it is defined by the violation of law—statutes that delineate improper conduct by fairly explicit criteria. Because some people break the law and others do not, populations can be dichotomized into law breakers (criminals) and law abiders (you and me). This circumstance encourages categorization of individuals into classes, from which it is then a short step to postulating discrete aetiologies accounting for the aberrant behaviours of those assigned consensus-defined forensic or clinical labels. Is it not also possible, however, that propensities to engage in behaviours that are likely to culminate in aggressive or criminal conduct reflect traits of continuous distribution in populations, but that only persons possessing such traits in a somewhat extreme degree (i.e. the relative end of a natural distribution of liability)—and abetted by environmental factors—ultimately meet criteria of clinical diagnosis or cross statutory thresholds of legal infraction? In this regard, it is interesting that virtually all of the psychiatric disorders associated with aggressive conduct also co-occur at rates exceeding chance. In turn, this extensive comorbidity has stimulated biometric analyses aimed at identifying shared and unique aetiologies of conduct-related disorders and candidate personality factors in broadly representative samples. This work supports a hierarchical model of alcohol and drug dependence, antisocial behaviour and disinhibitory temperament, whereby each component retains some specific genetic and/or environmental determinants. Yet just as noteworthy, all components also reflect the appreciable influence of a latent, highly heritable dimension of individual differences that is conducive to a spectrum of 'externalizing' disorders (see especially Krueger et al 2002). Such findings notwithstanding, debate over the relative merits of categorical and dimensional conceptualizations of psychopathology continue, although the conundrum of comorbidity and evidence that many clinical disorders share aetiologies with analogous personality traits may inform ongoing reformulations of psychiatric nosology (e.g. Kupfer et al 2002).

Brembs: You mentioned comorbidity with other factors. My take on this is one could define 'pathological' as follows. All the parameters that you look at are most likely continuous. If you draw a random sample from the whole population, it is very unlikely that an individual would end up at the far end on, say, three of those parameters. For such an individual, you would conclude that there is a pathology, is this correct? Are people who classify as pathological on this sort of definition the ones that commit the most crimes or who are the most conspicuous in society? Are these the ones we should focus on, or are there people that you wouldn't recognize on such a definition the ones that are most often responsible?

Manuck: I don't know. It is true, however, that a very small proportion of the population accounts for a large proportion of problematic behaviour. For instance, Walsh (2002) cites studies of two Philadelphia birth cohorts, 1945 (Wolfgang et al 1972) and 1958 (Tracy et al 1990). In the first, only 6% of boys accounted for over 70% of all homicides, aggravated assaults, rapes and robberies committed by cohort members. And in the second, 7.5% of boys committed over 60% of such crimes.

Hinde: The distinction between tendency to criminality and performing criminal acts is an important one. The thing that is left out of our discussion very often is the external stimulus. I don't think that Americans are on the whole worse than English people, but they do have guns, and that is one reason why the homicide rate is so much higher. The external stimulus is one of the important issues in triggering criminality into criminal acts.

Brembs: Let me try to rephrase the question. If you look at murder and then look at people who commit murders, would they classify as 'pathological' or as normal?

R Blanchard: What proportion of murders are family versus non-family? Are the family murders largely an expression of emotion and anger that are out of control? The non-family murders can be instrumental. A lot of murders in the USA involve kids who do have guns, who go into stores, get frightened and kill the person they are attempting to rob. It is not based on anger, but perhaps the activation of different neurobehavioural impulses. Murder is not a unitary dimension in terms of pathology, because of the variety and types of murder.

Martinez: In my opinion what is normal and not normal is decided by society at a certain time point. I would like to return to the example of wife battering. Now it is against the law in many countries, but it wasn't in the past. Some years ago all the men who battered their wives were perfect men, and they might have even been encouraged to do this if their wife was not obedient. Now we try to find some pathology in those men. However, in my opinion, it is not a matter of pathological aggression but of cultural norms.

Suomi: An analogy involves laws regarding smoking in the USA. A substantial proportion of the population, until 15 or 20 years ago, smoked. The proportion is now somewhat lowered, but the circumstances where that behaviour is acceptable have been changed dramatically. We now have 'no smoking' areas where anyone who tries to smoke in those areas is considered almost a criminal. The behaviour hasn't changed, but the way society has decided to define this situation has changed dramatically.

R Blanchard: So we have a social pathology in contrast to a psychopathology.

Suomi: The point is that what is acceptable is culturally determined to an extent. It may also vary developmentally. Richard Tremblay's data show that if what a normal two-year-old does in terms of proportion of acts that are aggressive were replicated in any older individual, it would absolutely be considered pathological.

Pfaff: I think we have been flirting with Björn Brembs' second question: what role does provocation play in aggression? And as a corollary, when is the provocative behaviour self aggressive? Would you like to elaborate on this?

Brembs: It was a question to the people who work in the field. We could also look at animal models. If we put an intruder into a mouse cage, we are provoking the aggression. But if an animal is moving into another's territory as a challenge, is this an aggressive act?

Martinez: In most cases of aggressive behaviour, provocation is needed, and there is an interaction involved. As I have already said, aggression is not a behaviour that occurs in a vacuum. Even when you commit suicide, it is not something that you are doing in isolation from society. You are committing suicide in relation to the people around you.

C Blanchard: The scenario we favour is one in which aggression is seen as an evolved biobehavioural system that is adaptive because it facilitates the control of resources. These can involve territory, status within a group, and access to females. The specific stimulus for aggression is a challenge from another conspecific for those kinds of resources. The provocation becomes very important. Perhaps we could put this back into a cultural context: pathological aggression is aggression occurring in response to an insufficient provocation. If the provocation is sufficient in terms of cultural norms, a proportionate aggressive response is not likely to be regarded as pathological.

Martinez: I completely agree that aggression is an adaptive behaviour in order to get resources. Imagine a group of children in a very poor neighbourhood. The only way they can do well is to be aggressive. It is a perfect behaviour for them to survive, but the same behaviour of these children in another context would be pathological, because it is not adaptive.

Lovell-Badge: I want to go through a list of topics I have picked up as we have been going along in our discussions that relate to neurogenesis. This is a topic that needs to be considered in the light of changes of behaviour, such as aggression. There are developmental issues. For example, you could have environmental or maternal effects on neurogenesis occurring as the embryo develops. Things could cross the placenta to trigger neurogenesis in a particular way. Randy Nelson commented on asymmetry in the head. There could also be asymmetry in the brain. There may be postnatal effects on neurogenesis. We heard from Steve Suomi about mothering. Are the effects of good mothering actually on gene expression, or are they on neurogenesis, increasing the numbers of serotonin-producing neurons? There are postnatal adult effects on neurogenesis. Neurogenesis does continue in the adult. It is generally thought in rodents to be restricted to certain areas of the brain, such as the lateral ventricles which contribute to the rostral migratory stream. Neurons from there contribute to the olfactory system and pheromone system. Factors that are affecting neurogenesis could have effects on those systems. The other main site where people are studying neurogenesis in the adult is the hippocampus and dentate gyrus. There are some data to suggest that particular types of learning and memory are affected by things that stimulate neurogenesis in that area.

Dulac: Also, stress inhibits this kind of neurogenesis.

Lovell-Badge: Yes, stress inhibits and exercise promotes it.

Pfaff: If you spend a lot of time on the treadmill, that means you have less time to be aggressive!

Lovell-Badge: There are also well known hormonal effects on neurogenesis. During pregnancy, could there be a role for neurogenesis in modifying behaviour to allow maternal aggression? There are likely to be seasonal effects on neurogenesis. We heard from Randy Nelson about the Siberian hamsters, where aggression increased during short day periods. Is this an effect on neurogenesis?

Nelson: That's a possibility, certainly. We see some seasonal changes in the hippocampus.

Suomi: We see seasonal effects on CSF 5-HIAA concentrations that might be related.

Lovell-Badge: We have seen exercise-induced neurogenesis. When one is using psychological ways to modify someone's behaviour, is it more likely to work if the person being treated is undergoing vigorous exercise? We need non-invasive ways of looking at neurogenesis so we can measure this. It is not always convenient to chop up your experimental animal.

Pfaff: Randy, you mentioned seasonality effects. Did you also want to say something about the possibility of CNS asymmetries in that study of yours?

Nelson: There is quite a literature on asymmetries in brain development and cases of brain injury, contributing to aggressive behaviour. Our study just addressed morphological asymmetries.

Lovell-Badge: We know that things can happen early in development which can have lifelong effects.

Pfaff: The biggest asymmetry I have read about within the scope of our discussion is in the frontal cortex. High levels of activity in the left prefrontal cortex are associated with a rosy view of life: the glass is half full. High activity in the right frontal cortex is associated with the opposite. Damage to either has opposite effects, such that damage to the left is associated with depression and damage to the right with exuberance.

Nelson: There are episodic differences in the relative degree of asymmetry, and then there are more chronic individual differences.

Pfaff: I want to address questions about human aggression more directly. Robert Hinde is going to begin this discussion.

Hinde: One thing needs to be said about the relation between maternal care and behaviour, and particularly subsequent aggression. If you have a dimension of maternal control (from harsh control to reasoned firm control to no control), and a dimension of maternal care, the children who are most aggressive lie not only in the area where there is very harsh control, but also in the permissive area where there is very little control, however much maternal care is given. The least aggressive children are in the area where there is reasoned control. This is something that should be taken on board in discussions of the relation between the effect of maternal care on subsequent expression of aggression.

Suomi: The most extreme form of control might be considered maltreatment, and the least control might be thought of as neglect. Both have pathological consequences.

Ferris: As we talk about the family dynamic and the outcome, certainly in the Rhesus studies the whole social structure is set up around the females. However, this is not the structure of the human family. Where do dads fit into this?

Hinde: There are very few data on the fathers, so I can't answer that. Fathers don't like being studied. It is a great problem.

There is one study in humans on a starvation situation in east Africa, where it was the difficult children who survived (De Vries 1984). This is relevant to what is pathological and what is adaptive. What is not pathological is fitting into the social norms, but being pathological may under some circumstances be a good thing for some individuals.

Pfaff: Catherine Dulac has a related point.

Dulac: I was intrigued by differences in the average levels of violence in societies, whether human or animals. I was fascinated by a recent study on baboons in which the most aggressive males in the group died from tuberculosis because they were the only ones able to access contaminated garbage cans. In the 10 years following this lack of alpha males, the group became incredibly less aggressive. In baboon societies which are terribly violent, subordinate males have a terrible fate. Their level of stress hormones is higher and their immune system is totally depressed. When the alpha males were eliminated for 10 years, suddenly that society became extremely peaceful. New males in the group adopted the peaceful mode of life (Sapolsky & Share 2004).

Suomi: There are some other interpretations. It may have been that some of the males were more readily accepted in the new group because they were peaceful. It is interesting, because there are both dramatic between species and within species differences, even with very closely related species. Carol Berman has a beautiful study in which she tracked maternal behaviour as a function of group size among the rhesus monkeys on Cayo Santiago island. What happens on that island is that social groups start off small and get bigger. Most groups grow from 30–40 individuals up to several hundred, and then each one splits up into smaller groups. She studied maternal behaviour throughout this process. She found that when the groups were small the mothers were laid back and let their kids have many interactions with non-family peers. The females who grew up representing the next generation then had a history of positive relations with one another. It was a nice, peaceful world. However, as each group got larger the proportion of non-kin to kin within a mother's immediate social space (an area of 5 m around her, where a mother can quickly retrieve her infant or address a direct attack, for example) increased. As this space started filling up more with non-family members, mothers started getting more restrictive and began discouraging their offsprings' non-family interactions. As a consequence, the next generation of females grew up with much fewer non-kin relationships and these were much less positive. The level of aggression in the group increased substantially. This trend accelerated to the point where these different families were literally at each other's throats most of the time. Eventually, the troops split up along family lines, and then the cycle went back to the beginning. This study demonstrates that mothers were responsive to demographic considerations in rearing their kids, and also that the long-term consequences of these alterations had implications for the integrity of the troop as a whole. Much of this was driven by aggressive behaviour in adults as a consequence of their differential rearing. There are also dramatic differences between macaque species in the relative levels of aggressiveness, the reciprocity of relationships, and the likelihood of reconciliation. Rhesus monkeys are off the scale in that they are really aggressive relative to a species like the Barbary macaque. These are species differences that may be related to some genetic differences.

Manuck: The social psychologist Richard Nisbett has conducted much interesting work on cultural differences in aggression and, in particular, what he refers to as 'cultures of honor' (Nisbett & Cohen 1996). He notes, for instance, that rates of homicide in the Southern and Southwestern states of the USA are higher than those in other regions of the country and that this is due to murders occasioned by interpersonal conflict, not homicide occurring in the commission of felonies. Then, in a series of elegant laboratory studies, he and his colleagues demonstrate that white males from the South show heightened sensitivity to cues of disrespect, threat, or challenge under mild experimental provocation. This is reflected in both behavioural and physiological signs of arousal and appears to spawn an increased readiness for confrontation violence.

Craig: An article by Zechner et al (2001) suggested that the X chromosome was accumulating genes to do with cognition and other behaviours in some kind of evolutionarily selective way. We know the *MAOA* gene is on there. As I mentioned previously, in undertaking twin studies we have always assumed that monozygous female twins have genetically identical genomes. However, it may be that their two X chromosomes are not equally inactivated. You may in fact get quite large amounts of skewing away from 50:50. The hypothesis based on this is that if you have QTLs affecting behaviour of any sort, this skewing could make them discordant for such traits. We thought we would look at correlations for behaviours between monozygotic (MZ) girl twins and MZ boy twins. We find that for most behaviours we studied MZ girl twins are more discordant than MZ boy twins, and this is especially marked for prosocial behaviour, peer relations and verbal ability. This suggests that there may be some QTLs for these traits on the X chromosome. The interesting thing is that if you look at DZ twins, because girls always share a paternal X, the other X they inherit can only change things a bit if there are QTLs there, therefore girl DZ twins should be more similar than male DZ twins. We find that exactly the same traits are significant in this way for the DZ situation. This suggests that there are some interesting genes on the X chromosome to do with the kinds of traits that we are interested in (Loat et al 2004).

Suomi: Wouldn't the same be true of siblings?

Craig: Yes, but they will have been exposed to a much greater degree to differential environments. In terms of the genetic potential that is right, though. The other aspect we wanted to look was the fact that there are genes on the human X chromosome that escape from inactivation. Several studies suggest that there are a lot more of them than is generally thought. They don't all have homologues on the Y, and therefore girls are going to have a higher expression of these genes than boys. We did a series of microarray experiments on lymphocytes. We find a range of genes that escape from inactivation in terms of significantly higher female to male ratio of expression. Several of these genes are expressed in brain and function in controlling other genes' activity.

Lesch: What regulates the escape?

Craig: The genes that escape tend to lie in clusters on the X chromosome. When we map them more accurately now we have the complete sequence of the X chromosome it turns out that some of them really are very close. There is also a purported relationship between the distribution of the genes that escape from activation and LINE elements.

Pfaff: Is there active demethylation?

Lovell-Badge: It is not clear how many of them escape from inactivation and how many of them are inactivated and then are released from this.

Craig: We are looking at the end product of a process. The other thing important to emphasize is that people have always thought of inactivation of being all or nothing. In fact, a lot of them are somewhat inactivated as opposed to totally inactivated.

Suomi: This is important, because you can have some modulation that yields a range of outcomes rather than an all-or-nothing effect.

Pfaff: You would agree that demethylation can be an active process? I ask once more because this has been a controversial question.

Craig: Yes, sure in the deprogramming and reprogramming cycle. When we were talking earlier about maternal effects, we were not far away from the idea that there can be methylation effects in response to behaviour.

Brodkin: In humans there are quite a large number of X-linked mental retardation syndromes (Frints et al 2002). This suggests that the X chromosome is rich in genes involved in brain development. Perhaps mutations in those genes of more severe effect might result in mental retardation, but perhaps other allelic variants of these genes (QTLs) on the X chromosome produce more subtle variations.

Keverne: If they escape inactivation but are imprinted you wouldn't pick up a difference in gene dosage. If you look at your data quantitatively, it might be possible to detect imprinted gene function.

Craig: We have been looking for an imprinted gene. We did pick one up, but it turned out to be *Xist*, which is what you'd expect.

Lovell-Badge: Methylation may not always be a good marker. In X-inactivation, the proteins that modify heterochromatin come on first, and then methylation is the last process.

Pfaff: I was hoping that Diane Robins would have a chance to talk more about androgen action. Can you give us a brief consideration?

Robins: I would emphasize that the androgen part is more important than the androgen receptor per se in terms of the behaviours, because of all the conversion of hormones (i.e. androgen being the precursor for oestrogen) and the potential for cross-regulation.

Pfaff: What I study in terms of the oestrogen receptors' functional genomics could be relevant to androgen action following aromatization.

Robins: Right, in that there is a lot of androgen that is converted to oestrogen.

Pfaff: You were also thinking about side-effects of various treatments, such as prostate cancer treatment.

Robins: As I mentioned, most people studying the androgen receptor at the molecular level now focus on the cancer aspects. One of the biggest problems is to get better anti-androgens, in part because of the side effects on male quality of life. It would be great to have things that are more tissue specific.

Pfaff: I was thinking about side effects of anabolic steroid use by athletes.

Martinez: Concerning androgens, what surprises me is that despite the huge amount of literature relating them to aggression in animals, this is hardly found in humans. At the present, the treatment with anti-androgens in humans is only used for reducing sexual aggression. It doesn't reduce aggression, but sexual drive.

Trainor: Most human studies on testosterone and aggression never take into account any social context. Measurements of behaviour are always done divorced from the time when any physiological measurement is taken. In studies I have looked at it is rare to see a close connection between behaviour and a hormone level.

Martinez: Most of the studies are correlation studies, and it is important to keep in mind that if you behave aggressively and are successful the level of testosterone will increase. There are very few causal studies on anabolic steroids in humans. There have been some controlled studies in the US army, and they show an increase in aggressiveness.

Robins: There is an interesting use of androgens occurring now in postmenopausal women. We could look for increased aggression there.

Pfaff: For the aggressive men, a suitable reference would be to the work of Harrison Pope from Harvard Medical School. In closing this meeting I'd like to thank all the participants for their contributions to the discussion.

References

Canteras NS, Ribeiro-Barbosa ER, Comoli E 2001 Tracing from the dorsal premammillary nucleus prosencephalic systems involved in the organization of innate fear responses. Neurosci Biobehav Rev 25:661–668

De Vries MW 1984 Temperament and infant mortality among the Masai of east Africa. Am J Psychiatry 141:1189–1194

Frints SG, Froyen G, Marynen P, Fryns JP 2002 X-linked mental retardation: vanishing boundaries between non-specific (MRX) and syndromic (MRXS) forms. Clin Genet 62:423–432

Krueger RF, Hicks BM, Patrick CJ, Carlson SR, Iacono WG, McGue M 2002 Etiologic connections among substance dependence, antisocial behaviour, and personality: modeling the externalizing spectrum. J Abnormal Psychol 111:411–424

Kupfer DJ, First MB, Regier DA 2002 A research agenda for DSM-V. Am Psychiatric Assoc, Washington DC

Loat CS, Asbury K, Galsworthy MJ, Plomin R, Craig IW 2004 X inactivation as a source of behavioural differences in monozygotic female twins. Twin Research 7:54–61

Nisbett RE, Cohen D 1996 Culture of honor: the psychology of violence in the South. Westview Press, Boulder, CO

Sapolsky RM, Share LJ et al 2004 A pacific culture among wild baboons: its emergence and transmission. PLoS Biol 2:E106

Tracy P, Wolfgang M, Figlio R 1990 Delinquency careers in two birth cohorts. Plenum, New York

Walsh A 2002 Biosocial criminology. Anderson Publishing, Cincinnati, OH

Wolfgang M, Figlio R, Sellin T 1972 Delinquency in a birth cohort. University of Chicago Press, Chicago

Zechner U, Wilda M, Kehrer-Sawatzki H et al 2001 A high density of X-linked genes for general cognitive ability: a run-away process shaping human evolution. Trends Genet 17:697–701

Index of contributors

Non-participating co-authors are indicated by asterisks. Entries in bold indicate papers; other entries refer to discussion contributions.

Subject index